Thomas Wiedl

Activity-based proteomics

Thomas Wiedl

Activity-based proteomics
Biomarker identification in lung cancer

Südwestdeutscher Verlag für Hochschulschriften

Impressum/Imprint (nur für Deutschland/only for Germany)
Bibliografische Information der Deutschen Nationalbibliothek: Die Deutsche Nationalbibliothek verzeichnet diese Publikation in der Deutschen Nationalbibliografie; detaillierte bibliografische Daten sind im Internet über http://dnb.d-nb.de abrufbar.
Alle in diesem Buch genannten Marken und Produktnamen unterliegen warenzeichen-, marken- oder patentrechtlichem Schutz bzw. sind Warenzeichen oder eingetragene Warenzeichen der jeweiligen Inhaber. Die Wiedergabe von Marken, Produktnamen, Gebrauchsnamen, Handelsnamen, Warenbezeichnungen u.s.w. in diesem Werk berechtigt auch ohne besondere Kennzeichnung nicht zu der Annahme, dass solche Namen im Sinne der Warenzeichen- und Markenschutzgesetzgebung als frei zu betrachten wären und daher von jedermann benutzt werden dürften.

Coverbild: www.ingimage.com

Verlag: Südwestdeutscher Verlag für Hochschulschriften GmbH & Co. KG
Heinrich-Böcking-Str. 6-8, 66121 Saarbrücken, Deutschland
Telefon +49 681 37 20 271-1, Telefax +49 681 37 20 271-0
Email: info@svh-verlag.de

Approved by: Switzerland, ETH Zurich, Diss., 2011

Herstellung in Deutschland:
Schaltungsdienst Lange o.H.G., Berlin
Books on Demand GmbH, Norderstedt
Reha GmbH, Saarbrücken
Amazon Distribution GmbH, Leipzig
ISBN: 978-3-8381-3137-5

Imprint (only for USA, GB)
Bibliographic information published by the Deutsche Nationalbibliothek: The Deutsche Nationalbibliothek lists this publication in the Deutsche Nationalbibliografie; detailed bibliographic data are available in the Internet at http://dnb.d-nb.de.
Any brand names and product names mentioned in this book are subject to trademark, brand or patent protection and are trademarks or registered trademarks of their respective holders. The use of brand names, product names, common names, trade names, product descriptions etc. even without a particular marking in this works is in no way to be construed to mean that such names may be regarded as unrestricted in respect of trademark and brand protection legislation and could thus be used by anyone.

Cover image: www.ingimage.com

Publisher: Südwestdeutscher Verlag für Hochschulschriften GmbH & Co. KG
Heinrich-Böcking-Str. 6-8, 66121 Saarbrücken, Germany
Phone +49 681 37 20 271-1, Fax +49 681 37 20 271-0
Email: info@svh-verlag.de

Printed in the U.S.A.
Printed in the U.K. by (see last page)
ISBN: 978-3-8381-3137-5

Copyright © 2012 by the author and Südwestdeutscher Verlag für Hochschulschriften GmbH & Co. KG and licensors
All rights reserved. Saarbrücken 2012

Preface

Lung cancer is the leading cause of all cancer related deaths worldwide. The 5-year survival rate ranges from 80% in early stages to less than 5% in advanced disease. Non-small cell lung carcinoma (NSCLC) accounts for 80% of all lung malignancies, 40% of which are adenocarcinomas (ACs) representing one of the most common lung cancer subtypes. Currently, prognosis is mostly determined based on the extension of disease at diagnosis. Thereby it has become evident that predicted and real outcomes can vary significantly, even for patients with the same stage of disease. Thus, novel biomarkers that refine prognosis with a reliable clinical significance are clearly needed.

In the genomic era biomarker studies typically aim at comparing transcript or protein abundances in malignant versus healthy tissues. However, taking into consideration that transcript and protein levels do not necessarily correlate with activity states, protein- and gene-expression profiling experiments might fail to detect changes in enzymatic activities caused by posttranslational events. Activity-based proteomics, a methodology that allows for the determination of enzymatic activity profiles of specific classes of enzymes even in complex proteomes, has become a valid option to circumvent this limitation. Based on the work of Prof. Cravatt, The Scripps Research Institute, USA and others an advanced activity-based proteomics platform for clinical biomarker discovery is presented in this PhD study. The implemented methodology was applied on 40 pairs of malignant and matching non-neoplastic human lung tissues to evaluate the potential of serine hydrolase (SH) activities, a large and diverse class of enzymes that have previously been linked to malignancies of the lung, as potential biomarker candidates for human lung adenocarcinoma.

The implemented methodology represents a label-free, solely mass spectrometry (MS) dependent platform for relative quantification of enzymatic activities in complex proteomes. Key features of this platform include good repeatability and reproducibility on a qualitative level as well as high quantitative precision. The results of this study revealed that the SHs TPSB2 protein, LYPLA1,

LYPLAL1, IAH1, ABHD10 and ABHD11 exhibited decreased or elevated activities in lung adenocarcinoma biopsies compared to matching non-neoplastic tissues in an almost exclusive manner, indicating that these enzymes play a general role in the pathogenesis of lung adenocarcinoma. On a transcript level, *ESD*, *TPBS2*, *IAH1* and *ABHD10* did not show any differences in lung cancer biopsies compared to normal tissues of the respiratory system as determined with a publicly accessible gene expression database. These observations indicate that the found SH activity differences are not detectable on a transcript level. Furthermore, the activities of ESD predicted the presence of high-grade tumors and the activities of ABHD11 predicted the development of distant metastases, both in a statistically significant model ($p<0.05$). Importantly, neither ESD nor ABHD11 have previously been associated with NSCLC. Although data validation with an increased sample set will be required to perform clinically more relevant statistical analysis, we anticipate that ESD and / or ABHD11 activities have the potential to develop into molecular predictors with a reliable clinical significance. We conclude that the implemented activity-based proteomics platform represents a powerful tool in the search for novel disease biomarkers and that SH activities bear predictive potential for human lung adenocarcinoma.

Table of contents

PREFACE	1
ABBREVIATIONS	5
1. INTRODUCTION	7
1.1 LUNG CANCER: CLASSIFICATION, EPIDEMIOLOGY AND PROGNOSIS	7
1.2 MOLECULAR EVOLUTION OF LUNG CANCER	9
1.3 LUNG CANCER AND THE BIOMARKER CONCEPT: STATUS QUO	11
1.4 MASS SPECTROMETRY (MS) BASED PROTEOMICS	13
1.5 ACTIVITY-BASED PROTEOMICS: A NOVEL STRATEGY FOR BIOMARKER DISCOVERY	16
1.6 THE SERINE HYDROLASE SUPERFAMILY	19
1.7 LUNG CANCER IN THE FUTURE	22
2. MATERIAL AND METHODS	23
2.1 SAMPLE COLLECTION	23
2.2 CELL CULTURE AND LYSIS	24
2.3 WESTERN BLOT	24
2.4 ACTIVITY-BASED PROBE	25
2.5 PROTEOME PREPARATION FOR LC-MS/MS ANALYSIS	26
2.6 LC-MS/MS ANALYSIS	28
2.7 DATABASE SEARCH AND LABEL-FREE QUANTIFICATION	29
2.8 STATISTICAL ANALYSIS	30
3. RESULTS	35
3.1 1D-SDS-PAGE DEPENDENT ACTIVITY-BASED PROTEOMICS	35
3.1.1 SH ACTIVITY PROFILES OF CALU-3 CELL EXTRACTS (MEMBRANE FRACTION)	35
3.1.2 SH ACTIVITY PROFILES OF CALU-3 CELL EXTRACTS (SOLUBLE FRACTION)	38
3.1.3 SH ACTIVITY PROFILES OF HUMAN LUNG ADENOCARCINOMA BIOPSIES	41
3.2 A MASS SPECTROMETRY (MS) DEPENDENT PLATFORM FOR ACTIVITY-BASED BIOMARKER DISCOVERY	43
3.2.1 CONSIDERATIONS FOR A SOLELY MS DEPENDENT ACTIVITY-BASED BIOMARKER DISCOVERY PLATFORM	43
3.2.2 ACTIVITY-BASED PROBES: FP-BIOTIN AND FP-1	44
3.2.3 DIRECTED IDENTIFICATION OF ACTIVE SHS IN COMPLEX PROTEOMES	46
3.2.4 LABEL-FREE QUANTIFICATION WITH *PROGENESIS LC-MS*	47
3.2.5 REPEATABILITY, REPRODUCIBILITY AND ROBUSTNESS	47
3.3 SH ACTIVITY PROFILES IN HUMAN LUNG ADENOCARCINOMA	51
3.4 BIOMARKER IDENTIFICATION	53
3.5 COMPARISON OF ACTIVITY- AND TRANSCRIPT-LEVELS OF ESD AND ABHD11	57
4. DISCUSSION	59
4.1 ACTIVITY-BASED PROTEOMIC STRATEGIES FOR THE INVESTIGATION OF SH ACTIVITIES IN COMPLEX PROTEOMES	59

  4.1.1 AN ADVANCED STRATEGY FOR THE IDENTIFICATION OF SH ACTIVITIES IN COMPLEX
  PROTEOMES 60
  4.1.2 LABEL-FREE QUANTIFICATION 62
  4.1.3 REPEATABILITY, REPRODUCIBILITY AND ROBUSTNESS 63
  4.1.4 ACTIVITY-BASED PROTEOMICS AND BIOMARKER DISCOVERY: TECHNICAL ASPECTS 65
4.2 SH ACTIVITIES AND HUMAN LUNG ADENOCARCINOMA **66**
  4.2.1 GLOBAL SH ACTIVITY PROFILES AS POTENTIAL BIOMARKERS FOR HUMAN LUNG
  ADENOCARCINOMA 66
  4.2.2 SH ACTIVITIES ASSOCIATED WITH HUMAN LUNG ADENOCARCINOMA 70
  4.2.3 IDENTIFICATION OF ESD AND ABHD11 ACTIVITIES AS BIOMARKER CANDIDATES FOR
  HUMAN LUNG ADENOCARCINOMA 73
  4.2.3.1 ESTERASE D (ESD) *73*
  4.2.3.2 ABHYDROLASE DOMAIN-CONTAINING PROTEIN 11 (ABHD11) *75*
  4.2.4 THE POTENTIAL OF ENZYMATIC ACTIVITIES AS CLINICAL BIOMARKERS: CHANCES AND
  LIMITATIONS 76

5. CONCLUSION **78**

SUPPLEMENTARY DATA **81**

PUBLICATIONS **97**

REFERENCES **98**

Abbreviations

^{19}F-NMR	Fluorine-19 nuclear magnetic resonance
1D-SDS-PAGE	One-dimensional sodium dodecyl sulfate polyacrylamide gel electrophoresis
ABP	Activity-based probe
ABPP	Activity-based protein profiling
AC	Adenocarcinoma
AUC	Area Under Curve
BSA	Bovine serum albumin
CAD	Collision activated dissociation
CID	Collision induced dissociation
Da	Dalton
DDA	Data dependent acquisition
DM	Distant metastasis
DMSO	Dimethylsulfoxide
DPBS	Dulbecco`s Phosphate Buffered Saline
EGFR	Epidermal growth factor receptor
ERCC1	Repair cross-complementation group 1
ESI	Electrospray ionization
FA	Formic acid
FDR	False discovery rate
FN	False negative
FP	False positive
FT-ICR	Fourier transform ion cyclotron
FWHM	Full width at half maximum
HPLC	High performance liquid chromatography system
IAA	Iodoacetamide
IAP	Inhibitor of apoptosis protein
ICAT	Isotope-coded affinity tags
LC	Liquid chromatography
LCC	Large cell carcinoma
LCM	Laser capture microdissection
LC-MS/MS	One dimensional liquid chromatography tandem mass spectrometry
LC/LC-MS/MS	Two dimensional liquid chromatography tandem mass spectrometry
LCNEC	Large cell neuroendocrine carcinoma
ln	Natural logarithm
LOH	Loss of heterozygosity
LOOCV	Leave-one-out cross-validation
LPA	Lysophosphatidic acid
LTQ	Linear trap quadrupole
m/z	Mass-to-charge ratio
MAGE	Monoalkylglycerol ether
MALDI	Matrix-assisted laser desorption/ionization
MDCK	Madin-Darby Canine Kidney
mg	Milligram

MMP	Matrix metalloproteinases
MS	Mass spectrometry
MS/MS	Tandem mass spectrometry
MudPIT	Multidimensional Protein Identification Technology
NSCLC	Non-small cell lung cancer
OCT	Optimal cutting temperature
PAF	Platelet-activating factor
PBS	Phosphate Buffered Saline
ppm	Parts per million
PVDF	Polyvinylidene fluoride
Rb	Retinoblastoma
ROC	Receiver Operating Characteristic
RP	Reversed-phase
rpm	Revolutions per minute
RPMI	Roswell Park Memorial Institute Medium 1640
R_T	Retention time
RT	Room temperature
SCC	Squamous cell carcinoma
SCLC	Small-cell lung cancer
SDS	Sodium dodecyl sulfate
SH	Serine hydrolase
SRM	Selected reaction monitoring
TCEP	Tris(2-carboxyethyl)phosphine hydrochlorid
TEV	Tobacco etch virus
TGF-β	Transforming growth factor beta
TKI	Tyrosine kinase inhibitor
TN	True negative
TP	True positive
UC	Ulcerative colitis
VEGF	Vascular endothelial growth factor

1. Introduction

1.1 Lung cancer: classification, epidemiology and prognosis

Lung cancer is the major cause of all cancer related deaths with a worldwide mortality of 1.2 million each year (Parkin et al., 1999; Garcia, 2007). Lung cancer is a mostly smoking related disease with a time lag of more than 20 years between exposure and the onset of symptoms (Haugen, 2008; Proctor, 2001). Dyspnea, pain and cough are typical symptoms manifested by patients with lung cancer and dependent on localization of the tumor, locoregional spread and effects of metastatic growth (Sculier, 2008). Unfortunately, since early lung cancer is rarely symptomatic, most patients present in advanced disease with metastases having already spread to other organs (Blum, 2005).

Lung cancer can be divided into non-small cell lung cancer (NSCLC, comprising 80% of all lung cancers) which is thought to originate from lung epithelial cells and small-cell lung cancer (SCLC, accounting for 20% of all lung cancers) which is a tumor of neural crest origin (Sharma et al., 2007). Thereby, it is important to mention that SCLC is the main subtye of the group of neuroendocrine tumors of the lung that comprises typical carcinoid, atypical carcinoid, large cell neuroendocrine carcinoma (LCNEC) and SCLC (Chong et al., 2006). However, NSCLC comprises the three major histotypes squamous cell carcinoma (SCC), large cell carcinoma (LCC) and adenocarcinoma (AC), the latter representing 30% of all lung cancers (Brambilla, 2008; Chen et al., 2003; Kaira et al., 2008). The five-year survival rate following surgical treatment for NSCLC ranges from 80% in early stages to less than 5% in inoperable advanced disease, with overall 5-year survival rates being comparable in the three NSCLC subtypes (Korst, 2008; Hong, 2008; Minna, 2005).

Incidence rates for lung cancer have been increasing in Europe since the beginning of the 20th century for men and women until the 1980s and 1990s, respectively (Janssen-Heijnen and Coebergh, 2003). In Europe, decreasing incidence rates for SCC and LCC and increasing incidence rates for AC have been reported, the latter representing the most frequent lung cancer subtype in the United

States (Janssen-Heijnen and Coebergh, 2003). Interestingly, AC represents the most frequent lung cancer subtype among never-smokers (Sun et al., 2007). However, several studies suggest that the introduction of filter cigarettes in the 1950s and a change in smoking behavior might account for increasing AC incidence rates (Janssen-Heijnen and Coebergh, 2003). The rationale for this hypothesis is, that smokers take deeper puffs to compensate for lower nicotine yields which in turn leads to a more intensive penetration with carcinogenic substances of lung zones where ACs typically develop (Janssen-Heijnen and Coebergh, 2003).

However, assessment of tumor size, lymph node status and the presence of metastases at diagnosis, a process referred to as "staging" (TNM classification of malignant tumors) is currently the method of choice for determining prognosis and treatment modality for patients suffering from lung cancer (Dusmet, 2008; Mullon, 2008). This classification system, administered by the International Union Against Cancer (UICC) aims to assess the extend of the disease, whereby the "T" descriptor indicates the size of the primary tumor, the "N" descriptor the extent of lymph node involvement and the "M" descriptor describes the presence or absence of distant metastases (Dusmet, 2008). For each of the descriptors, numeric subscripts indicate how far the disease has already progressed (Dusmet, 2008). TNM subsets with similar prognosis for survival and with similar treatment options are combined into "stage groups" (0, IA, IB, IIA, IIB, IIIA, IIIB, IV) (Dusmet, 2008). Thereby, it has become evident that predicted and real outcomes can vary significantly in individuals, also for patients with the same stage of disease (Selvaggi, 2008). For example: surgical resection remains the method of choice for patients suffering from stage I and II NSCLC and according to current clinical treatment guidelines, no adjuvant therapy, i.e. postoperative chemotherapeutic treatment, is indicated for stage I NSCLC patients (Scott et al., 2007; Korst, 2008). Approximately 25% to 30% of patients with NSCLC have stage I disease (Raponi et al., 2006). Out of these patients only 50% to 65% will not relapse within 5 years after surgical treatment (Raponi et al., 2006). Therefore, for the remaining 35% to 50% of patients an adjuvant therapeutic intervention might have been beneficial. Currently there is no characteristic of any kind available to reliably stratify patients into groups that are at high risk for relapse.

1.2 Molecular evolution of lung cancer

Environmental factors like tobacco smoke and genetic susceptibility interact to influence lung cancer development, whereby hormonal and viral factors unrelated to smoking have also been suggested (Yano et al., 2006; Herbst et al., 2008). As a consequence of tobacco smoke exposure, tissue injury occurs in the form of genetic and epigenetic changes (mutations, loss of heterozygosity (LOH) and promoter methylation) (Herbst et al., 2008). These changes can persist over a long period of time and may lead to aberrant pathway activation and dysregulation of cell growth and apoptosis, two hallmarks of cancer that consequently may result in premalignant changes like dysplasia and clonal patches (Hanahan and Weinberg, 2000; Herbst et al., 2008; Mao et al., 1997). In more detail, early events in the development of NSCLC include LOH at the chromosomal regions 17p13 (encoding for p53), 9p21 (encoding for p16^{INK4a}) and 3p21.3 (encoding for Ras association domain-containing protein 1), all of which are tumor suppressor genes (Wistuba et al., 2002). Additionally, mutations in the epidermal growth factor receptor (*EGFR*) gene as well as in the V-Ki-ras2 Kirsten rat sarcoma viral oncogene homolog (*KRAS*) represent early events in the development of lung adenocarcinoma (Westra, 2000; Tang et al., 2005).

A large number of molecular alterations have been identified in NSCLC, some of which are discussed in the following: ErbB1 (also known as EGFR) and ErbB2, another member of the EGFR family, are commonly overexpressed in NSCLC (Selvaggi et al., 2004; Poulsen, 2008). The *MYC* proto-oncogene is frequently amplified in NSCLC (Broers et al., 1993). Although the exact role of *MYC* amplification in lung cancer pathogenesis still remains to be elucidated, recent studies suggest that the *MYC* gene product promotes cell cycle progression by activating key molecules responsible for entry into the S-phase of the cell cycle (Poulsen, 2008). Survivin, an inhibitor of apoptosis protein (IAP) is frequently overexpressed in NSCLC and it has been shown that absence of survivin correlates with improved prognosis (Poulsen, 2008). The *TP53* tumor suppressor gene is the most commonly mutated gene in cancer (Tennis et al., 2006). The p53 encoding gene *TP53* is mutated in 50% to 70% of all lung adenocarcinomas (Herbst et al., 2008). Genetic alterations range from LOH as indicated above over homozygous deletions and DNA rearrangements to point mutations with every event contributing to the inactivation of p53 (Iggo et al., 1990). Inactivation of p53 leads to a loss of control over the G1 arrest and subsequently to uncontrolled cell growth (Brambilla et al., 2003). In 34% of all lung adenocarcinoma cases the serine / threonine kinase 11 (LKB1) encoding gene *STK11* is mutated, whereby it has been shown that LKB1 acts as a mediator of p53-dependent cell death (Karuman et al., 2001). The tumor suppressor protein LKB1 has furthermore been shown to modulate lung cancer differentiation and metastases and LKB1 loss has been associated with increased susceptibility to 2-deoxyglucose treatment for NSCLC (Ji et al., 2007; Inge et al., 2009). The *CDKN2A* gene encoding for the p16^{INK4a} protein is frequently inactivated in NSCLC, resulting in reduced activity of the retinoblastoma (Rb) tumor suppressor and it has been proposed that inactivation of the p16^{INK4a} / Rb pathway plays a mandatory role in the pathogenesis of lung cancer (Otterson et al., 1994; Poulsen, 2008). The p16^{INK4a} gene locus is located at 9p21, a region that is frequently subjected to LOH in NSCLC (Merlo et al., 1994). Within the same locus, the protein p14 is encoded which indirectly inhibits the degradation of p53 (Poulsen, 2008). The transforming growth factor β (TGF-β) signaling network, best known for its role in cell cycle inhibition, has also been shown to be altered in lung cancer (Poulsen, 2008). In 70% - 100% of all NSCLC cases, mutations in chromosome 3p have been reported (Kok et al., 1987). Importantly, a number of tumor suppressor candidate genes like *FHIT* and *RASSF1A* are encoded in this region (Wistuba et al., 2000; Dammann et al., 2000). Finally, the expression of the angiogenesis promoting vascular endothelial growth factor (VEGF) has been shown to be upregulated in lung cancer and clinical trials with the VEGF targeting antibody bevacizumab have resulted in statistically significant survival benefits for patients suffering from NSCLC (Yano et al., 2006).

However, despite the substantial insights that have been gained during the past decades regarding the molecular pathogenesis of lung cancer, only a few findings have been translated into the clinics (Poulsen, 2008). The role of *EGFR* mutations and excision repair cross-complementation group 1 (ERCC1) expression is discussed in the following (Paez et al., 2004; Olaussen et al., 2006).

1.3 Lung cancer and the biomarker concept: status quo

According to the Biomarkers Definitions Working Group, a biomarker is defined as "a characteristic that is objectively measured and evaluated as an indicator of normal biological processes, pathogenic processes, or pharmacologic responses to a therapeutic intervention" (Atkinson et al., 2001).

In the genomic era scientists typically refer to molecular characteristics, i.e. transcript and protein abundances as well as genetic mutations as biomarkers. Basically, three types of biomarkers can be distinguished: prognostic, predictive and pharmacodynamic biomarkers (Sawyers, 2008). Prognostic biomarkers predict the outcome of an individual cancer without therapeutic intervention and are used to guide the decision which patient to treat or which patient might benefit from additional therapeutic interventions (Sawyers, 2008). Predictive biomarkers on the other hand are employed to assess whether a given patient will benefit from a particular treatment (Sawyers, 2008). For example, a subset of NSCLC patients harbor somatic activating mutations in the kinase domain of the *EGFR* gene and respond well to the tyrosine kinase inhibitors (TKIs) gefitinib (marketed as Iressa by AstraZeneca, Teva) and erlotinib (marketed as Tarceva by OSI Pharmaceuticals, Genentech and Roche) (Sharma et al., 2007). Thus, these activating mutations represent predictive biomarkers. Finally, pharmacodynamic biomarkers measure the treatment effects of a drug on the tumor and can be used to fine-tune drug doses at an early stage in clinical development (Sharma et al., 2007). Therefore, pharmacodynamic biomarkers can be employed to find drug doses that are required to induce tumor shrinkage but still lie below the level of toxicity (Sharma et al., 2007).

Currently, the majority of NSCLC biomarkers available are per definition predictive. Probably the most prominent amongst them being activating mutations in the kinase domain of the *EGFR* gene (Herbst et al., 2008). In 2004, several research groups independently identified mutations in the kinase domain of *EGFR* in approximately 10% of specimens derived from patients with lung adenocarcinoma in the United States and in 30% to 50% of biopsies from Asian patients suffering from adenocarcinoma of the lung (Sequist et al., 2007). It has been shown that these mutations occur most frequently in Asian women that have never smoked and as mentioned above, it has been demonstrated that patients harboring these mutations respond well to treatments with the tyrosine kinase inhibitors gefitinib and erlotinib (Paez et al., 2004; Pao et al., 2004). More than 80% of the *EGFR* mutations occurring in lung cancer represent in-frame deletions within exon 19 or the substitution of arginine for leucine at position 858 (L858R) (Herbst et al., 2008; Sharma et al., 2007). However, it should be mentioned that many patients that initially respond well to the TKIs gefitinib and erlotinib, eventually relapse (Pao et al., 2005). Studies have demonstrated that a substitution of methionine for threonine at position 790 (T790M) in the kinase domain (exon 20) of *EGFR* is associated with acquired resistance (Pao et al., 2005; Sharma et al., 2007). Amplification of the *MET* proto-oncogene has been shown to be another major mechanism of acquired resistance to TKIs in patients suffering from lung adenocarcinoma (Bean et al., 2007).

A second example for a predictive biomarker for NSCLC is the expression of the protein ERCC1 that plays a central role in the nucleotide excision repair pathway that recognizes and removes covalently linked cisplatin-DNA adducts (Mu et al., 1996; Zamble et al., 1996). It has been shown that patients with completely resected NSCLC and ERCC1 negative tumors as determined by immunohistochemistry benefit from adjuvant cisplatin-based chemotherapy in comparison to patients with ERCC1 positive tumors (Olaussen et al., 2006).

In summary, somatic mutations within the *EGFR* gene as well as reduced expression levels of ERCC1 represent molecular biomarkers with a clinically relevant predictive significance. The establishment of these biomarkers in the clinics represents an important step forward towards the goal of individualized medicine were patients are selected for specific treatment regimens based on molecular characteristics peculiar to each tumor and each individual subject (Selvaggi, 2008). However, relevant *EGFR* mutations occur only in a small fraction of patients suffering from NSCLC in the Western world and although ERCC1 expression can help refining current treatment regimens, no evidence has so far been presented that ERCC1 represents a molecular target that can be exploited for the development of novel treatment approaches for NSCLC (Sequist et al., 2007).

In order to contribute to the long term aim of personalized medicine for patients suffering from lung cancer, the predictive potential of enzymatic activities in human lung adenocarcinoma was investigated in this study for two reasons: first, little is known about the prognostic and / or predictive potential of enzymatic activities as biomarkers in human lung cancer. Second, enzymes are generally thought of as "druggable" biomolecules and thereby represent potential targets for novel therapeutic approaches (Hopkins and Groom, 2002; Russ and Lampel, 2005). The methodology employed for the determination of enzymatic activity profiles in human specimens was activity-based proteomics and is discussed in detail in section 1.5.

1.4 Mass spectrometry (MS) based proteomics

Per definition, proteomics refers to the investigation of proteins on a large scale with particular focus on protein structures, their interactions and functions (Blackstock and Weir, 1999; Anderson and Anderson, 1998). For qualitative and partially quantitative analysis of proteins in complex proteomes, mass spectrometry has increasingly become the methodology of choice (Aebersold and Mann, 2003). The principle of mass spectrometry is the detection of ionized analytes based on their molecular weights whereby the mass spectrometer itself consists of an ion source that produces ionized analytes, a mass analyzer that measures the mass-to-charge (m/z) ratio and a detector that counts the number of ions at a given m/z value (Aebersold and Mann, 2003). Most notably, the rapid development of mass spectrometry based proteomics was accelerated by the discovery and development of protein ionization methods, namely electrospray ionization (ESI) and matrix-assisted laser desorption / ionization (MALDI), for which the Nobel prize in chemistry was awarded in 2002 (Aebersold and Mann, 2003). Four basic mass analyzers are currently used in mass spectrometry based proteomics: ion trap, time-of-flight (TOF), quadrupole and Fourier transform ion cyclotron (FT-ICR) mass analyzers (Aebersold and Mann, 2003).

Currently, the most common implementation for protein biomarker discovery studies relying on mass spectrometry is referred to as "shotgun proteomics" (one dimensional liquid chromatography tandem mass spectrometry (LC-MS/MS) is typically employed in shotgun proteomics based measurements) (Hu et al., 2007; Wolters et al., 2001; Washburn et al., 2001). In such an experiment, a complex protein mixture is digested prior to mass spectrometric analysis, cleaved peptides are separated on a reversed-phase (RP) high performance liquid chromatography system (HPLC, in the context of mass spectrometry based proteomics typically abbreviated with "LC" for liquid chromatography) and finally analyzed on an instrument equipped with a nanoelectrospray ion source (Hu et al., 2007; Wolters et al., 2001; Washburn et al., 2001). Mass spectrometers commonly used for shotgun proteomics are linear trap quadrupole-Orbitrap (LTQ-Orbitrap) or LTQ Fourier transform ion cyclotron resonance (LTQ-FTICR, also referred to as LTQ-FTMS) instruments, the latter exhibiting extraordinary sensitivity, mass accuracy, resolution and dynamic range (Aebersold and Mann, 2003).

In a discovery-driven shotgun experiment, ionized analytes (referred to as precursor ions) are selected automatically based on signal intensities derived from a survey scan (also referred to as MS scan), a process called data dependent acquisition (DDA) (Aebersold and Mann, 2003). Thereby, the most abundant ions are accumulated in the ion trap and fragmented through collision induced dissociation (CID), a fragmentation mechanism that is also referred to as collision activated dissociation (CAD) (Aebersold and Mann, 2003). Fragments are then analyzed by the second mass analyzer (MS/MS scan) and signals are finally recorded by the detector (Hu et al., 2005). In order to identify proteins, several computational algorithms have been implemented that compare observed peptide fragment masses with theoretical peptide fragment masses derived from databases containing comprehensive protein sequence information (Aebersold and Mann, 2003). Hence, only proteins with primary sequences available are identified with this approach.

Besides protein identification, several strategies for quantification can be employed in shotgun proteomics: first, it has been shown that the total number of MS/MS spectra recorded for a given peptide fragment correlates with relative protein abundance, a method referred to as spectral counting (Liu et al., 2004). Second, several strategies for accurate quantification relying on stable isotopes have been developed were quantification is based on the comparison of precursor ion currents derived from heavy and light versions of isotopically labeled analytes (Domon and Aebersold, 2010; Gygi et al., 1999a). Third, in a strategy referred to as label-free quantification, precursor ion currents are compared between different samples measured under rigorously controlled instrument conditions without the use of stable isotopes (Domon and Aebersold, 2010). By making use of advanced software solutions, precursor ion patterns derived from many experiments can be overlaid and compared (Domon and Aebersold, 2010).

It is important to mention that a mass spectrometry based technology referred to as selected reaction monitoring (SRM) or "targeted" proteomics has recently gained popularity in the field of proteomics due to its capability of reliably quantifying even low abundant analytes in complex mixtures and the large number of peptides that can be quantified in a single LC-MS experiment (Lange et al., 2008). In an SRM experiment, a predefined precursor ion and one of its fragments are selected by two mass filters of a triple quadrupole instrument and monitored over time for precise quantification (Lange et al., 2008). However, in order to conduct SRM-based experiments the exact masses of precursor / fragment ion pairs need to be available (Lange et al., 2008). For that reason, the SRMAtlas database containing comprehensive and validated information on transitions for the model organism yeast has been established (Picotti et al., 2008). As announced on the SRMAtlas website (*mrmatlas.org*) in January 2011, the database will soon be expanded by transitions for the vast majority of proteins (95%) constituting the human proteome.

However, in discovery-driven biomarker studies where protein abundances in normal versus disease states are compared, no information on protein identities is typically available prior to analysis. Until comprehensive information on transitions for peptides matching to proteins constituting the proteome of interest becomes available for SRM-based experiments, shotgun proteomics remains the methodology of choice for mass spectrometry based biomarker discovery studies (Hung and Yu, 2010). Furthermore, for shotgun proteomics it is possible, based on information derived from experiments in DDA mode, to create an inventory of detected peptides that are of potential interest in so-called inclusion lists (Domon and Aebersold, 2010). Similarly, an inventory of peptides matching to contaminants or proteins that are of no interest to the researcher can be created and stored in exclusion lists. Both lists can then directly be imported into the mass spectrometer operating software. This strategy, often referred to as "directed" mass spectrometry, has significant advantages over conventional shotgun experiments in DDA mode since the bias in favor of the most intense signals is partially removed which in turn provides a deeper penetration into the investigated proteome (Domon and Aebersold, 2010). In this study, we employed a directed strategy with the mass spectrometer operating in DDA mode only when no signals according to the inclusion list were detected in the survey scan, thereby combining the advantages of experiments in DDA mode and directed proteomics. The strategy employed in this study is illustrated in Figure 1.

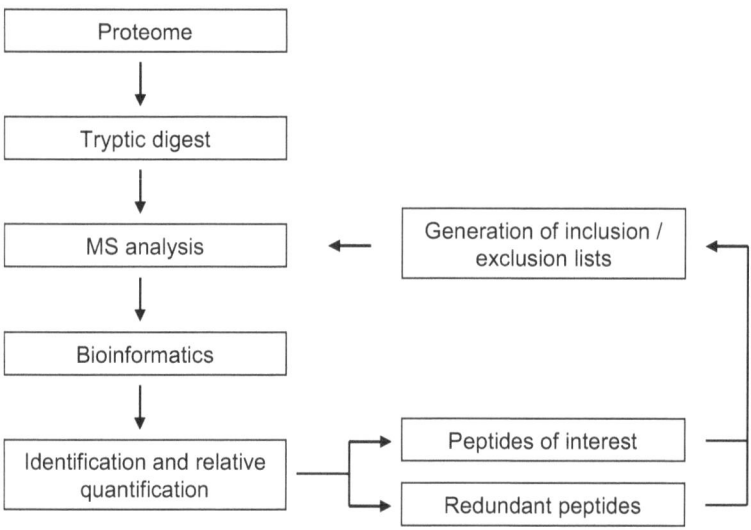

Figure 1 - A directed strategy for the identification and label-free relative quantification of proteins in complex proteomes. The proteome of interest is subjected to tryptic digestion and the resulting peptide mixture is analyzed using mass spectrometry. Several computational algorithms can be employed for protein identification and label-free quantification. Information on peptides matching to proteins that are of special interest, i.e. differentially expressed proteins or members of a predefined protein family, as well as information on peptides matching to redundant proteins can be extracted and information stored in so-called inclusion and exclusion lists for subsequent directed analyses (Domon and Aebersold, 2010).

1.5 Activity-based proteomics: a novel strategy for biomarker discovery

Over the past decades substantial insights have been gained into the molecular pathogenesis of malignancies in general and of NSCLC in particular (Sawyers, 2008; Rodriguez-Pineiro et al., 2010; Poulsen, 2008). However, moderate progress has been made in the translation of this knowledge into clinical biomarkers with a reliable predictive significance (Sawyers, 2008; Rodriguez-Pineiro et al., 2010).

One explanation for this unsatisfactory situation is that biomarker discovery studies typically compare transcript or protein abundances in normal versus disease states. Considering that transcript and corresponding protein levels do not necessarily correlate with activity states, protein- and gene-expression profiling methods might fail to detect crucial changes in enzymatic activities caused by posttranslational events during tumor progression and treatment response (Gygi et al., 1999b; Alaiya et al., 2000; Sieber and Cravatt, 2006; Sieber et al., 2006; Jessani et al., 2005). Activity-based proteomics, also referred to as activity-based protein profiling (ABPP) has become a promising option to circumvent this limitation. In summary, active-site directed chemical structures, so-called activity-based probes (ABPs) are employed to covalently target active enzymes. Activity-based probes typically consist of two key elements: a reactive group that covalently binds to the active site of members of a distinct enzyme family and a reporter tag used for identification and / or quantification of targeted enzymes. The reporter tag is either a fluorophor (rhodamine, for example) that is used for visualization of targeted enzymes by one-dimensional sodium dodecyl sulfate polyacrylamide gel electrophoresis (1D-SDS-PAGE) or biotin which is employed for enrichment of labeled enzymes and subsequent identification and / or quantification by MS (see Figure 2) (Nomura et al., 2010a). Since inactive enzymes remain untagged, this strategy represents a valid approach to determine enzymatic activity profiles of specific classes of enzymes even in complex proteomes (Liu et al., 1999; Jessani et al., 2002).

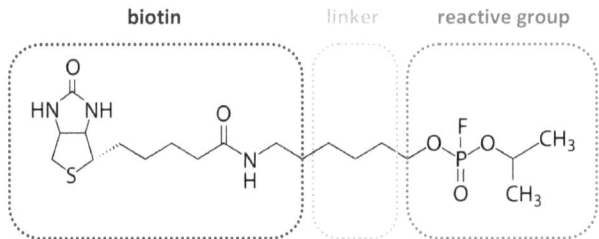

Figure 2 - Structure of the activity-based probe "FP-1" employed in this project (see also section 2.4). An activity-based probe typically consists of a reactive group, a non-reactive linker region and a reporter tag, biotin in case of FP-1.

Activity-based proteomics has already been introduced as a high-throughput platform for the analysis of primary human specimens (Jessani et al., 2005). Thereby a quantitative, high-throughput and 1D-SDS-PAGE dependent analysis step is followed by a qualitative and highly sensitive MS-based phase (two dimensional liquid chromatography tandem mass spectrometry (LC/LC-MS/MS), also referred to as Multidimensional Protein Identification Technology (MudPIT)) in which representative samples are analyzed (Jessani et al., 2005).

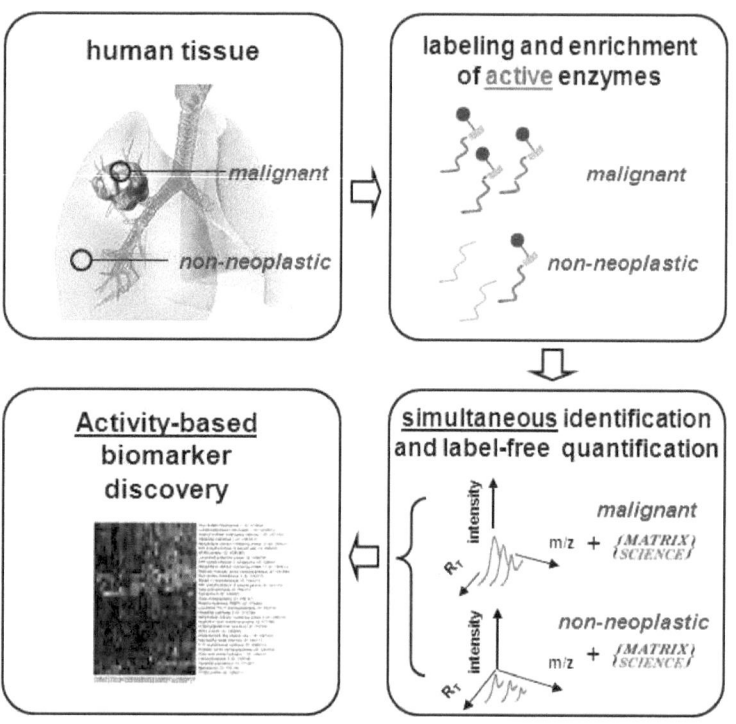

Figure 3 - Workflow for activity-based biomarker discovery implemented in this study. Malignant and matching non-neoplastic tissues were labeled with the ABP FP1 (see section 2.4) and after MS analysis, label-free quantification was performed by making use of the software package *Progenesis LC-MS* (see section 2.7).

However, it has become evident that the depth of the analysis is restricted by the dynamic range inherent to 1D-SDS-PAGE (see also section 3.1). Therefore, an alternative strategy with an optimized trade-off between throughput and sensitivity is desirable for clinical activity-based biomarker discovery studies. Making use of existing activity-based proteomics platforms and recent advances in software solutions for label-free quantification we implemented a directed, semi-quantitative and solely LC-MS/MS dependent workflow for the investigation of serine hydrolase activities in human lung adenocarcinoma (see Figure 3) (Jessani et al., 2005; Jessani et al., 2002).

1.6 The serine hydrolase superfamily

The serine hydrolase (SH) superfamily comprises a large and diverse repertoire of enzymes that make up 1% of the human proteome (Jessani et al., 2005). Members of the serine hydrolase superfamily have extensively been studied in an activity-based manner, most likely due to the unique molecular characteristics of their active site, the so-called catalytic triad (Simon and Cravatt, 2010). The catalytic triad is an alignment of three amino acids that is conserved within members of the SH superfamily. This triad consists of a serine, a histidine and a glutamate or aspartate residue (Cai et al., 2004; Ekici et al., 2008). In the proposed mechanism for the inhibition of active lipases, a subgroup of the serine hydrolase superfamily, through phosphonates as suggested by Dijkstra et al., the phosphorous atom of the phosphonate group binds to the nucleophilic oxygen of the serine residue in the active site of serine hydrolases (see Figure 4) (Dijkstra et al., 2008). Since fluorophosphonate / fluorophosphate derivatives react irreversibly and stoichiometrically with the active-site serine residue, these chemical structures can be used as powerful probes to isolate and identify active serine hydrolases (Dijkstra et al., 2008; Liu et al., 1999).

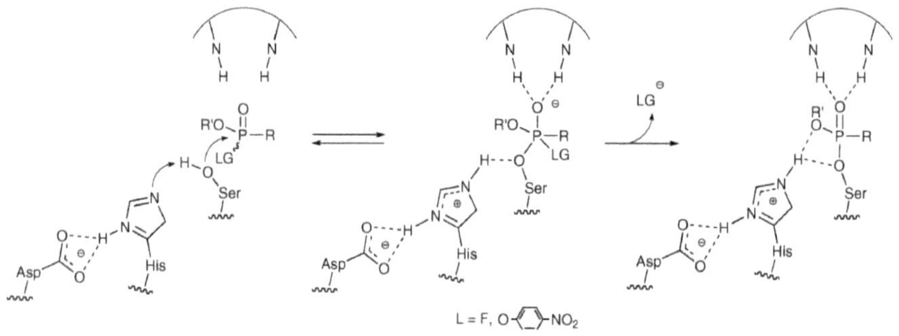

Figure 4 - Proposed mechanism for the inhibition of lipases by reactive phosphonates (Dijkstra et al., 2008). The reaction mechanism presented above illustrates the stoichiometric nature of the covalent inhibition exploited in activity-based proteomics.

Members of the SH superfamily have previously been linked to cancer (Kuhajda, 2000; Chen and Kelly, 2003; Huang et al., 2004) and lung cancer (Meyer et al., 2004; Zelvyte et al., 2004). Seprase (UniProtKB / Swiss-Prot ID: Q12884) for example, is a cell surface serine protease that has been shown to promote tumor growth in mouse models of human breast cancer (Huang et al., 2004). The serine protease Cathepsin G (UniProtKB / Swiss-Prot ID: P08311) is thought to influence tumor cell invasiveness by contributing to the activation of matrix metalloproteinases (MMPs) in lung cancer (Zelvyte et al., 2004). By employing activity-based proteomics, the previously uncharacterized protein Arylacetamide deacetylase-like 1 (UniProtKB / Swiss-Prot ID: Q6PIU2) has been shown to exhibit elevated activities in aggressive human breast cancer specimens (Jessani et al., 2005). Another example of a SH that has previously been linked to cancer is Fatty acid synthase (FASN, UniProtKB / Swiss-Prot ID: P49327), a large and multifunctional protein that is involved in the synthesis of palmitic acid (Chakravarty et al., 2004). This enzyme contains a SH subunit and has been shown to be expressed at high levels in human malignancies of the breast, colon, ovary and lung (Chakravarty et al., 2004; Kuhajda, 2000; Orita et al., 2007). FASN is seen as an attractive new target for novel NSCLC treatments since it has been reported that inhibition of FASN significantly inhibits the growth of orthotopic xenograft tumors from human NSCLC cell lines in mice (Orita et al., 2007). Another SH, namely human carboxylesterase (UniProtKB / Swiss-Prot ID: P23141), has been shown to convert the anti-NSCLC agent irinotecan into its metabolically active form, thereby increasing cytocidal effects 100-fold (Ohtsuka et al., 2003).

The enzymes discussed above have been chosen as representative examples of the SH superfamily, because all mentioned enzymes have been identified within this project in human lung adenocarcinoma biopsies and matching non-neoplastic tissues. Furthermore, in a recently published study by Liu et al., annexin A3, a SH inhibitor, has been identified as a novel biomarker candidate for lung adenocarcinoma (Liu et al., 2009b). It is worth mentioning that annexin 3 is a phospholipase inhibitor, whereby several members of this group of enzymes were identified in human lung adenocarcinoma biopsies within this study (Supplemental Table S2) (Cunningham et al., 2005). Despite these findings it is noteworthy that one of the most commonly used biomarkers in oncology, prostate specific antigen (PSA, UniProtKB / Swiss-Prot ID: P07288), is a member of the serine hydrolase superfamily (Balk et al., 2003; Hernandez and Thompson, 2004).

Finally, as indicated above, in an activity-based proteomics study the previously uncharacterized protein Arylacetamide deacetylase-like 1 (UniProtKB / Swiss-Prot ID: Q6PIU2) has been identified to exhibit activities at a high level in aggressive human breast cancer specimens (Jessani et al., 2005). By making use of competitive activity-based proteomics, i.e. the incubation of a given proteome with an activity-based probe in presence of a candidate small molecule inhibitor, Leung et al. were able to identify and Chiang et al. further developed a potent and selective inhibitor targeting Arylacetamide deacetylase-like 1 (Leung et al., 2003; Chiang et al., 2006). With untargeted LC-MS analysis of lipophilic metabolites derived from cancer cells that have been incubated with the specific inhibitor for Arylacetamide deacetylase-like 1, Chiang et al. found that Arylacetamide deacetylase-like 1 is the principal 2-acetyl monoalkylglycerol ether (2-acetyl MAGE) hydrolase (Chiang et al., 2006). A reduction in Arylacetamide deacetylase-like 1 activities led to a reduction in MAGE levels *in vitro* which in turn was found to perturb other pro-tumorigenic lipids like alkyl-lysophosphatidic acid (alkyl-LPA) (Chiang et al., 2006). These changes were associated with reductions in migration and *in vivo* tumor growth indicating an important role of Arylacetamide deacetylase-like 1 in the pathogenesis of cancer (Chiang et al., 2006; Nomura et al., 2010a).

Based on the scientific evidence presented above, we believe that serine hydrolases play an important role in the pathogenesis of NSCLC and hypothesize that serine hydrolase activities bear predictive potential for human lung adenocarcinoma. In order to test this hypothesis an advanced, activity-based biomarker discovery platform with an optimized trade-off between throughput and sensitivity has been implemented in this study and serine hydrolase activity profiles derived from 40 pairs of human lung adenocarcinoma biopsies and corresponding normal lung tissues were linked to clinical follow-up data.

1.7 Lung cancer in the future

Global cigarette consumption peaked in the 1990s but due to the 35-year time lag between consumption and mortality, the World Health Organization (WHO) calculated that lung cancer rates will increase until 2020 or 2030 (Proctor, 2001; WHO, 1999). Furthermore, it is expected that the global burden of lung cancer will undergo substantial changes in the near future from the developed to the developing world with over 300 million new smokers in China (Selvaggi, 2008). For this reason, the tobacco epidemics seems to be unfolding in China and India like in Europe and the United States 40 years ago (Proctor, 2001).

Novel treatments and molecular biomarkers with a reliable predictive significance will play major roles in individualized medicine were patients suffering from lung cancer are selected for treatment regimens based on molecular characteristics specific to each tumor and each individual subject (Selvaggi, 2008). Nonetheless, despite the enormous efforts that are undertaken to improve early detection and treatment options, lung cancer remains a mostly smoking related disease and prevention should therefore be of highest socio-economic priority (Haugen, 2008; Selvaggi, 2008).

2. Material and Methods

2.1 Sample collection

Forty patients with lung adenocarcinoma underwent surgical resection at the University Hospital Zurich (UHZ) between 2003 and 2006. Tissue samples from tumor and matching non-neoplastic lung parenchyma were snap-frozen and stored at -80°C (tumor tissue: 208mg +/- 70mg and non-neoplastic tissue: 188mg +/- 94mg; data presented as mean weight +/- standard deviation). Pathological staging was conducted according to the 6th TNM classification of malignant tumors (2002). An overview of patient characteristics is given in Table 1, for detailed information see Supplemental Table S1. Quality control for tumor content per surface and tumor viability was performed on whole sections of frozen samples. Malignant biopsies were therefore embedded in an optimal cutting temperature (OCT) compound (Tissue Tek), sections of 8 micrometres (μm) thickness were manufactured with a HM 560 cryostat (Microm), hematoxylin and eosin (H&E) stained and morphologically reviewed by a pathologist from the UHZ (Supplemental Table S1) (Steu et al., 2008). Human subjects participating in the study have given the requisite informed consent.

Characteristics	N = 40
Sex - no. (%)	
Male	25 (62)
Female	15 (38)
Age - year	
Mean	66.5 +/- 11.0
Range	41.1 - 86.1
Smoking status - no. (%)	
Never smoker	5 (13)
Previous / current smoker	35 (87)
Clinical stage - no. (%)	
IA	6 (15)
IB	11 (28)
IIA	2 (5)
IIB	3 (8)
IIIA	11 (28)
IIIB	2 (5)
IV	5 (13)
Type of surgery - no. (%)	
pneumonectomy	30 (75)
lobectomy	7 (17.5)
wedge resection	3 (7.5)
Lymph node metastases - no. (%)	
N0	18 (45)
N1	8 (20)
N2	12 (30)
N3	2 (5)
Grade[1] - no. (%)	
G1	4 (10)
G2	14 (36)
G3	21 (54)

[1] Grade was not available for one patient.

Table 1 - Clinical characteristics of study participants.

2.2 Cell culture and lysis

Roswell Park Memorial Institute Medium 1640 (RPMI 1640, Biowest) containing 10% fetal bovine serum (FBS, Biowest) and 50µg/ml gentamicin sulfate (Biowest) was used for culturing CaLu-3 cells, a human lung adenocarcinoma cell line (Haws et al., 1994). CaLu-3 cells were obtained from frozen stocks stored in media containing the supplements described above and additionally 10% dimethylsulfoxide (DMSO, Fluka). After the frozen cell suspension has carefully been thawed, cells were plated in 10ml RPMI medium containing the supplements mentioned above without DMSO and incubated at 37°C until cells reached 90% confluence. Media change was performed every second day. To prepare cell extracts, cells were washed 3 x with 10ml ice cold Dulbecco`s Phosphate Buffered Saline (DPBS, Biowest) and without making use of trypsin a rubber policeman was employed to detach cells from culturing plates. The cell suspension was centrifuged in an Eppenddorf 5810 R centrifuge for 5min at 4°C at 3000 revolutions per minute (rpm), the cell pellet resuspended in 500µl DPBS and the suspension transferred to a tissue grinder (Kontes Glass Co., size: 22) and agitated manually 100 times. After dounce homogenization, cell suspensions were sonicated (60% output, 80% duty cycle, 20 times, Heat Systems Ultrasonics, Inc.) and centrifuged for 5min at 4°C at 3000rpm to remove debris. When indicated, cell lysates were separated into soluble and membrane fractions by ultracentrifugation for 45min at 4°C at 64000rpm with an Optima™ TLX Personal Benchtop Ultracentrifuge (Beckman Coulter). Total extracted protein concentration was determined using the *DC* Protein Assay kit (BioRad) and finally adjusted to 1mg/ml with Phosphate Buffered Saline (PBS). For 1D-SDS-PAGE dependent, activity-based experiments, activity-based probe (FP-rhodamine, see section 2.4) labeling of 50µg total protein extracts derived from CaLu-3 cells as described above or from human biopsies (see section 2.5) was carried out by incubating proteomes for 1hr at room temperature (RT) using 2µM FP-rhodamine final concentration (FC). Fluorescent signals were subsequently detected with a Typhoon 9400 scanner (Amersham).

2.3 Western Blot

Electrophoretic separation of isolated proteins was performed using 10% precast acrylamide gels (BioRad). 10µg of cell extracts were diluted each with 4 x sample buffer (8% (w/v) sodium dodecyl sulfate (SDS, Sigma), 40% (w/v) glycerol (Sigma), 0.04% (w/v) bromophenol blue (Roth) and 0.5% (v/v) β-mercaptoethanol (Sigma) in 1M Tris-HCl (pH = 6.8)) to a final volume (FV) of 30µl, heated at 90°C for 8min and loaded into respective wells of the precast gel. After electrophoretic separation, proteins were transferred to a polyvinylidene fluoride membrane (PVDF, BioRad) using a wet transfer system (BioRad). Immunodetection was carried out after the membrane has been blocked with 5% (w/v) bovine serum albumin (BSA, Sigma) diluted in Tris-Buffered Saline containing 0.05% (w/v) Tween (TBS-T, Sigma) for 1hr at RT by employing antibodies against actin (mouse anti-human, MP-Biomedicals) dissolved 1:10000 in TBS-T containing 5% (w/v) BSA for primary incubation (1hr at RT). After that membranes were incubated for 1hr at RT with a rabbit anti-mouse horseradish peroxidase conjugate (Sigma) dissolved 1:10000 in TBS-T containing 5% (w/v) BSA and chemiluminescent signals were detected using a LAS-3000 imaging system (Fujifilm).

2.4 Activity-based probe

Three activity-based probes (ABPs) were employed in this study: the fluorophosphates FP-rhodamine (see Figure 5) and FP-biotin (see Figure 6) as well as the fluorophosphonate derivative "FP-1" (see Figure 7). Fluorophosphate / fluorophosphonate derivatives inhibit the majority of SHs irreversibly in a covalent manner, whereas other hydrolases or inactive SHs remain unlabeled (Liu et al., 1999; Creighton, 1993; Walsh, 1979). FP-rhodamine has successfully been employed for targeting active SHs in complex proteomes and subsequent visualization on 1D-SDS-PAGE (Liu et al., 1999; Jessani et al., 2005). As shown in Figure 5, the fluorophore 5-carboxytetramethylrhodamine (5-TAMRA) serves as a reporter tag for the activity-based probe FP-rhodamine. Both, FP-biotin and FP-1 have been developed and successfully employed in several studies for the inhibition and isolation of active SHs (Liu et al., 1999; Higson et al., 1999; Nomura et al., 2005; Quistad and Casida, 2004). FP-rhodamine as well as FP-biotin were gifts from Prof. B. Cravatt, The Scripps Research Institute, La Jolla, USA. FP-1 was purchased from Toronto Research Chemicals (TRC) and remained stable upon reconstitution (5mM in DMSO) during the whole course of analysis as monitored by fluorine-19 nuclear magnetic resonance (^{19}F-NMR, see section 3.2.2).

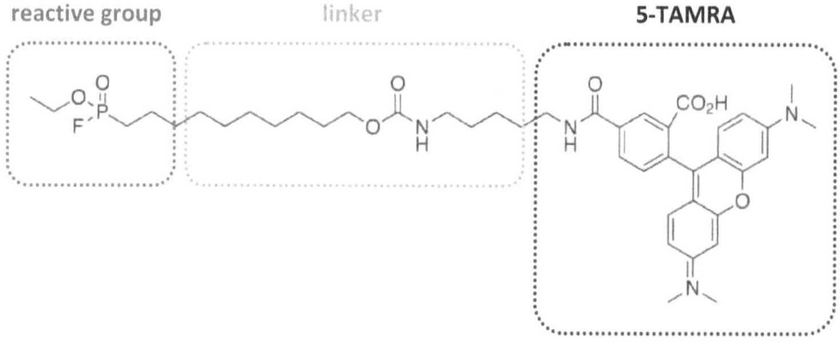

Figure 5 - Structure of the activity-based probe FP-rhodamine. The fluorophore 5-carboxytetramethylrhodamine (5-TAMRA) serves as a reporter tag (Jessani et al., 2005).

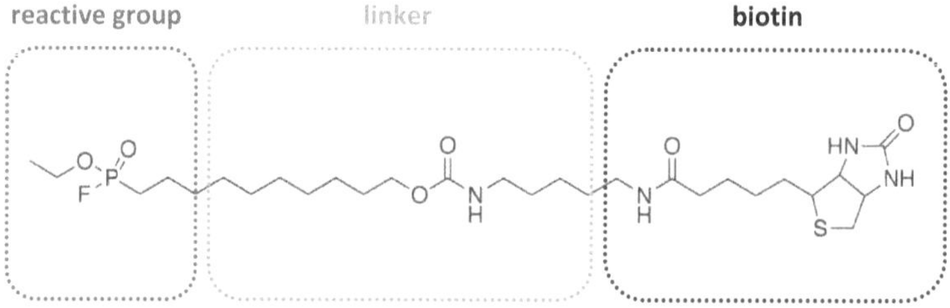

Figure 6 - Structure of the activity-based probe FP-biotin. The linker region between the reactive group and biotin differs in size compared to the linker region in FP-1 (see Figure 7) (Jessani et al., 2005).

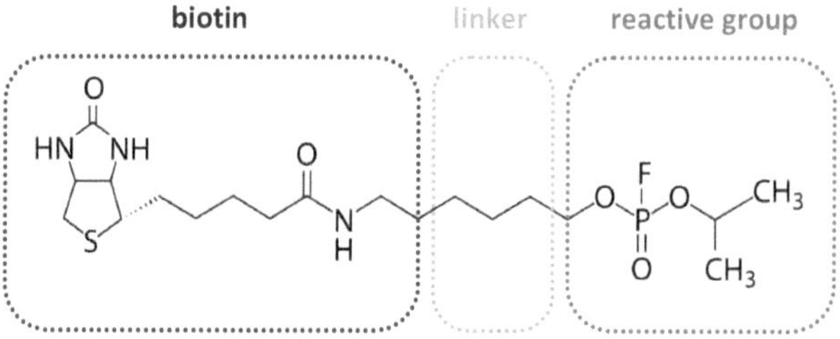

Figure 7 - Structure of the activity-based probe FP-1. The nucleophilic serine-OH in the active site of SHs attacks the phosphorus atom of the reactive fluorophosphate group. The stoichiometrical reaction is irreversible and binding occurs a covalent manner (Dijkstra et al., 2008).

2.5 Proteome preparation for LC-MS/MS analysis

Human tissues (~150mg) were dounce homogenized in 500µl PBS (pH = 7.4), sonicated (60% output, 80% duty cycle, 20 times, Heat Systems Ultrasonics, Inc.)

and centrifuged for 5min at 4°C at 3000rpm to remove debris. Total extracted protein concentration was determined using the *DC* Protein Assay kit (BioRad) and adjusted to 1mg/ml with PBS.

Activity-based probe (FP-biotin or FP-1) labeling of 1mg total protein extract derived from human biopsies as described above or from CaLu-3 cells (see section 2.2) was carried out by incubating proteomes for 2hrs at RT using 5µM FP-biotin / FP-1 FC. Proteomes were solubilized by adding Triton X-100 (Sigma) to 1% FC (v/v) followed by 1hr rotation at 4°C. Samples were then desalted (PD MidiTrap G-25, GE Healthcare), SDS (Sigma) was added to a FC of 0.5% (w/v) and activity-based probe labeled proteomes were incubated for 8min at 90°C. Next, samples were combined with 67.5µl of pre-washed streptavidin coated agarose beads (Pierce), adjusted to 8.5ml FV with PBS and rotated for 1hr at RT to allow binding of FP-biotin / FP-1 labeled enzymes to streptavidin coated agarose beads. After binding, beads were washed twice with 1% SDS (w/v), twice with 6M urea and three times with PBS (each wash rotating for 5min at RT at 8.5ml FV followed by centrifugation for 3min at 1400rpm at 4°C). Enriched samples were then reduced for 1hr with tris(2-carboxyethyl)phosphine hydrochlorid (TCEP, Sigma) at 5mM FC and alkylated for 1hr with iodoacetamide (IAA, Sigma) at 10mM FC. Next, beads were washed once with 1ml 50mM ammonium carbonate ($(NH_4)_2CO_3$), reconstituted in 400µl 50mM $(NH_4)_2CO_3$ and 0.75µg trypsin (Promega) were added to the solution followed by 12hr incubation at 37°C. After that, samples were centrifuged for 10min at 14000rpm at 4°C and formic acid (FA) was added to the peptide containing supernatant to 5% FC (v/v). Samples were stored at -80°C. Prior to MS analysis, samples were purified with C-18 Ultra MicroSpin columns (The Nest Group, Inc.) and peptide concentrations were measured using the Qubit (Invitrogen) quantification platform.

For every biopsy, 3 proteome aliquots were independently incubated with FP-biotin / FP-1 (= biochemical triplicates), affinity purified and subsequently analyzed twice with MS (= technical duplicates). Proteomes processed as described above without FP-biotin / FP-1 incubation served as negative controls. For all biopsies, sample quality was assessed by 1D-SDS-PAGE (3µg of proteome were loaded per well of 10% precast acrylamide gels, BioRad) and subsequent Coomassie Blue (BioRad) staining (see Supplemental Figure S1). FP-biotin / FP-1 labeling efficiencies were determined for all samples through parallel incubation of 1µg of the SH trypsin (Promega) with FP-biotin / FP-1 at 5µM FC in 100µl PBS (see Supplemental Figure S1).

2.6 LC-MS/MS analysis

Tryptic peptide solutions were analyzed on an Eksigent-Nano-HPLC system (Eksigent Technologies) linked to a linear ion trap FT-ICR hybrid mass spectrometer (LTQ-FTMS, Thermo Finnigan) equipped with a nanoelectrospray ion source (Thermo Scientific). Peptide solutions were loaded on self-made tip columns (75µm inner diameter, 80mm length) containing reversed phase C18 material (AQ, 3µm particle size, 200 angstrom (Å) pore size, Bischoff GmbH). Peptides were eluted at a flow rate of 200nl/min over a 60min gradient (0 - 50min: 3% to 30% solvent B, 50 - 58min: 30% to 45% solvent B, 58 - 60min: 45% - 97% solvent B). Solvent composition was 0.2% FA (v/v), 1% acetonitrile (ACN) (v/v) for solvent A and 0.2% FA (v/v), 80% ACN for solvent B.

For initial experiments in data dependent acquisition (DDA) mode, one high resolution survey scan acquired in the ICR cell was followed by three CID MS/MS scans in the linear ion trap of the three most intense signals of the survey scan. For one high resolution survey scan, 10E+6 ions were accumulated over a maximum time of 500ms and full width at half maximum (FWHM) resolution was set to 100000 at 400m/z. Signals exceeding 500 ion counts triggered an MS/MS attempt and 10E+4 ions were accumulated for an MS/MS scan over a maximum time of 300ms. Charge state screening was enabled and singly and unassigned charge states were rejected. Precursor masses already selected for MS/MS measurements were excluded for further selection for 20s (50 parts per million (ppm) exclusion window). To assess LC-MS/MS performance, control samples containing 400 attomole (amol) tryptic peptides of the glycoprotein Fetuin A (UniProtKB / Swiss-Prot ID: P12763, Sigma) were frequently analyzed during the whole course of experiments.

After initial MS experiments in DDA mode, inclusion and exclusion lists were generated for directed sequencing. Inclusion lists contained m/z values and corresponding retention times (R_T) of tryptic peptides matching to SHs observed in initial experiments (R_T window = observed R_T +/- 5min). Further m/z values of peptides matching to SHs were added to these lists utilizing the publicly accessible database *PeptideAtlas* (Desiere et al., 2005). Exclusion lists contained equivalent information for peptides matching to streptavidin (Pierce) and trypsin (Promega). Inclusion / exclusion lists (.txt file format) were imported directly into the global mass lists of the MS operating software Xcalibur version 2.0.7 (Thermo Fisher Scientific, Inc.). The m/z tolerance for inclusion and exclusion lists was set to +/- 10ppm. When no signals according to inclusion lists were detected, the nth most intense feature of the survey scan was selected for the nth MS/MS attempt in DDA mode.

2.7 Database search and label-free quantification

To assess MS analysis performance, raw data (.raw file format) were converted into .mzXML files and visualized on a two-dimensional density plot utilizing the software tool *Pep3D* (Li et al., 2004). Raw data of acceptable quality were loaded into the commercial software package *Progenesis LC-MS* version 2.6 (Nonlinear Dynamics), a software tool developed for label-free quantification of LC-MS data. Data analysis was performed for each patient individually. LC-MS data were normalized and aligned according to manufacturers specifications. Mascot generic files (.mgf file format) generated with *Progenesis LC-MS* were searched against a human protein database (European Bioinformatics Institute (EBI), ftp.ebi.ac.uk/pub/databases/SPproteomes/fasta/proteomes/, downloaded on February 11, 2009) using the Mascot 2.2 search engine. Parameters for precursor tolerance and fragment ion tolerance were set to +/- 5ppm and +/- 0.8 Dalton (Da), respectively. Using a decoy (reversed) database, false discovery rates (FDRs) were calculated and Mascot ion score cut-offs set according to FDR < 5% (protein level) and FDR < 1% (peptide level). Proteins detected in negative controls and enzymes not belonging to the SH superfamily as determined by literature search and *in silico* sequence comparison were excluded from further analysis (Marchler-Bauer et al., 2009; Marchler-Bauer et al., 2007; Marchler-Bauer et al., 2002).

For quantification, we adapted a strategy that utilizes the median ratio of the top three intense peptide ions matching to a protein to relatively quantify enzymes in malignant versus matching non-neoplastic tissues (Grossmann et al., 2010;

Malmstrom et al., 2009; Silva et al., 2006). Only peptides matching the following criteria were used for quantification: i) peptide sequences uniquely identified proteins in the human organism and did not share sequence similarities with trypsin (Promega) or streptavidin (Pierce). Peptides matching to different proteins or to protein isoforms that originated from the same gene location as determined by Ensembl Genome Browser (*www.ensembl.org/Homo_sapiens/Info/Index*) were used for quantification. ii) Peptides needed to be detected in every replicate analyzed for a given patient. iii) No missed cleavages in peptide sequences were allowed. iv) No methionine in peptide sequences was allowed. v) Peptide charge: +2 or +3. vi) Minimum normalized ion abundance (from here on referred to as "relative intensity") as calculated by *Progenesis LC-MS*: 2.5E+02. In exceptional cases the following rules applied: if only one peptide matched to a protein, activity ratios were calculated through direct division of the two respective relative intensity values. If two peptides matched to a protein, the average activity ratio calculated from the four respective relative intensities was reported. In rare cases, peptides exceeded the minimum relative ion intensity of 2.5E+02 in all replicates of a given biopsy, but were not detected at all in replicates of the corresponding tissue. In these cases activity ratios were set to 5 or -5 (log2 transformed), respectively.

2.8 Statistical analysis

Two-sided, unpaired Student`s t-test (*R* version 2.8.1) was employed to evaluate validity of the implemented quantification strategy with data derived from label-free LC-MS/MS experiments and subsequent analysis with *Progenesis LC-MS* version 2.6 (Nonlinear Dynamics) as described in section 2.7.

In order to link clinical data with SH activities, between-group comparisons of normally distributed data were performed with two-sided, unpaired Student`s t-test (*R* version 2.8.1). Between-group comparisons of non-normally distributed data were conducted by using the Mann–Whitney U test (*R* version 2.8.1). For enzymatic activities showing statistically significant differences according to clinical parameters as calculated with Student`s t-test or Mann–Whitney U test, logistic regression was employed to evaluate the prediction abilities of the respective enzymatic activities.

Logistic regression is a statistical methodology that is especially well suited to be employed in statistical data analysis with dichotomous outcomes, for example diseased / healthy (Peng et al., 2002). To be more precise, logistic regression is a valid model for hypotheses testing about relationships between a categorical outcome variable (in case of this study a dichotomous outcome, i.e. development of distant metastases: yes / no; presence of high-grade tumors: yes / no) and one or more categorical or continuous predictor variables (in case of this study: one continuous predictor variable, i.e. one SH activity) (Peng et al., 2002). Since observation-to-predictor ratios of 10 to 1 have been reported for logistic regression in the literature, we only focused on one predictor variable per time for model building (Peng et al., 2002). Due to the categorical nature of the variables, linear regression is not well suited to be employed for the analysis of such datasets. One solution to analyze such data with linear regression is the creation of a plot where categorized predictor variables are plotted against the mean values of the outcome variables of the respective predictor categories (Peng et al., 2002). The resulting curve will be of sigmoidal shape, i.e. S-shaped (Peng et al., 2002). However, due to mathematical reasons, this solution is suboptimal (Peng et al., 2002). Logistic regression circumvents these limitations by applying the so-called logit function to the outcome variable (Peng et al., 2002). Thereby, the logit of the outcome variable is the natural logarithm (ln) of the odds of the outcome variable, whereby the odds is the ratio of the probability that the outcome variable will happen (i.e. the probability that a patient will develop distant metastases at a given enzymatic activity value) to the probability that the outcome variable will not happen (i.e. the probability that a patient will not develop distant metastases at a given enzymatic activity value) (Peng et al., 2002).

After a significant relationship between a dichotomous outcome variable and a SH activity has been detected with logistic regression (*R* code see below), leave-one-out cross-validation (LOOCV) was employed. With LOOCV it is possible to test whether the logistic regression model built on enzymatic activity values derived from human biopsies investigated in this study (= training data) can be generalized by employing the model on a validation data set. The principle of LOOCV is discussed in the following: the first observation from a given data set is removed, a model is then built based on the remaining observations (= training data) and the model is tested on the observation that has initially been removed (= validation data). After that, the second observation of the complete data set is removed, a model is built based on the remaining data and the model is tested again on the observation that has been removed. This algorithm is repeated for every observation of the data set.

Based on the results obtained with LOOCV, misclassification rates (see Equation 1; TP = number of true positives, FP = number of false positives, TN = number of true negatives, FN = number of false negatives), sensitivity (see Equation 2) and specificity (see Equation 3) were calculated and the optimal trade-off between sensitivity and specificity was assessed by Receiver Operating Characteristic (ROC) analysis. Area Under Curve (AUC) values were calculated using the trapezoidal rule (Loong, 2003).

$$Misclassification-rate = \frac{FP+FN}{TP+FP+TN+FN} \qquad \text{Equation 1}$$

$$Sensitivity = \frac{TP}{TP+FN} \qquad \text{Equation 2}$$

$$Specificity = \frac{TN}{TN+FP} \qquad \text{Equation 3}$$

The *R* code used in this project for hypotheses testing with logistic regression and for cluster analysis (see section 3.3 and Figure 16) is given below.

```
################################################################
#                                                              #
#              L O G I S T I C   R E G R E S S I O N           #
#                                                              #
################################################################

# R script adapted from http://cran.r-project.org/ and
# http://yatani.jp/HCIstats/LogisticRegression

x<-read.table(file.choose(), header=TRUE)
attach(x)

y<-t(x)
```

```r
write.csv(y, file='/Desktop/B.csv')

res<-glm(GRADE~E1, family=binomial(logit), data=x)
summary(res)

# manual LOOCV

x<-read.table(file.choose(),header=TRUE)
attach(x)

res<-glm(GRADE~E1, family=binomial(logit), data=x)
summary(res)

a<- res$coeff[1]
b<- res$coeff[2]

z<-3.6218
p<-(exp(a+b*z))/(1+exp(a+b*z))
p

# Predict rate calculation

predict_activity <- predict(res, type="response")
predict_activity

# Plotting of predict rates

plot(E21, GRADE, cex=1.5)
points(E21, predict_activity, pch=3, col="red", cex=1.5)
v <- seq(-15, 15, 0.1)

# y <- exp(INTERCEPT + COEFFICIENT*v) / (1 + exp(INTERCEPT +
# COEFFICIENT*v))

a<- -1.1689
b<- 1.0143
y<-(exp(a+b*v))/(1+exp(a+b*v))
lines(v, y, col="blue", lwd=2)

# Probability calculation

z<-2.5
p<-(exp(a+b*z))/(1+exp(a+b*z))
p

# ODDS
p<-(exp(a+b*z))/(1+exp(a+b*z))
ODDS<-p /(1-p)
```

```
###############################################################
#                                                             #
#                    H E A T M A P                            #
#                                                             #
###############################################################

# R script adapted from http://cran.r-project.org/,
# http://www2.warwick.ac.uk/fac/sci/moac/students/peter_cock/
# r/heatmap/ and
# http://manuals.bioinformatics.ucr.edu/home/R_BioCondManual

# packages required: "gplots", "gdata" and "gtools"

library(gplots)

# LOAD DATA

a<-read.table(file.choose(), header=TRUE)

# REWRITE IN NUMERIC MATRIX

b<-as.matrix(a)

# use t() to obtain transposed matrix
c<-t(b)

# MINIMAL VALUE OF MATRIX

min<-min(b)

# MAXIMAL VALUE OF MATRIX

max<-max(b)

# ADJUST INTERVAL FOR COLOR KEY

mycol <- colorpanel(n=999,low="red",mid="black",high="green")
pairs.breaks <- c(seq(-4.7, 0, length.out=500),seq(0, 5.3,
length.out=500))

# PLOT HEATMAP

heatmap.2(b, breaks=pairs.breaks, col=mycol, key=TRUE, keysize
= 1, lmat=rbind( c(0, 3, 4), c(2,1,0) ), lwid=c(1,5,2),
lhei=c(1, 4), density.info="none", dendrogram="column",
cexRow=0.95, cexCol=0.9, trace="none", scale="none",
na.color="darkblue", margin=c(7,9))
```

3. Results

3.1 1D-SDS-PAGE dependent activity-based proteomics

Here we describe the implementation of a 1D-SDS-PAGE dependent activity-based proteomics workflow for the investigation of SH activities in complex proteomes. The presented SH activity profiles were derived from proteomes of the human lung adenocarcinoma cell line CaLu-3 and of human lung adenocarcinoma biopsies and corresponding normal lung tissues by employing the activity-based probe FP-rhodamine (see section 2.4).

3.1.1 SH activity profiles of CaLu-3 cell extracts (membrane fraction)

Cell extracts were prepared as described in section 2.2. Standard labeling conditions as outlined in section 2.2 were adapted from Jessani et al. (Jessani et al., 2005). In order to determine optimal labeling conditions, the loading amount of CaLu-3 derived proteomes, FP-rhodamine concentration or both were sequentially increased.

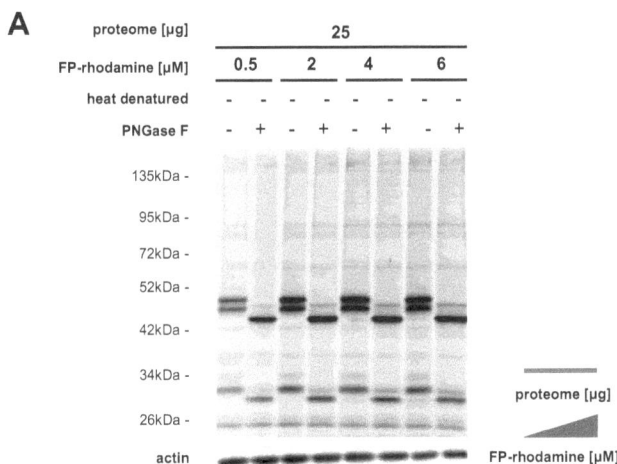

Figure 8A - SH activity profiles of CaLu-3 cell extracts (membrane fraction) at increasing FP-rhodamine concentrations. Proteome incubation at 4µM or 6µM FP-rhodamine did not result in a higher number of detected SH activities compared to incubation at 2µM FP-rhodamine. PNGase F was used for protein deglycosylation. Actin served as a loading control.

Incubation of cell extracts at standard labeling conditions (1mg/ml proteome at 2µM FP-rhodamine) resulted in the detection of approximately 10 distinct bands that correspond directly to enzymatic activities (see Figure 8A). As illustrated in Figure 8A, the incubation of proteomes at increasing FP-rhodamine labeling concentrations in comparison to standard labeling conditions does not result in the detection of additional SH activities in case of CaLu-3 derived membranous cell extracts.

Figure 8B shows detected SH activities at increasing CaLu-3 proteome loading amounts incubated at 2µM FP-rhodamine. This experiment revealed that in comparison to standard labeling conditions, a further increase in proteome loading amounts does not lead to the detection of additional SH activities. Taking into consideration that proteomes derived from human biopsies are typically of limited quantity, it was decided to use 25µg as a loading amount for future 1D-SDS-PAGE dependent activity-based experiments for membrane fractions derived from human biopsies.

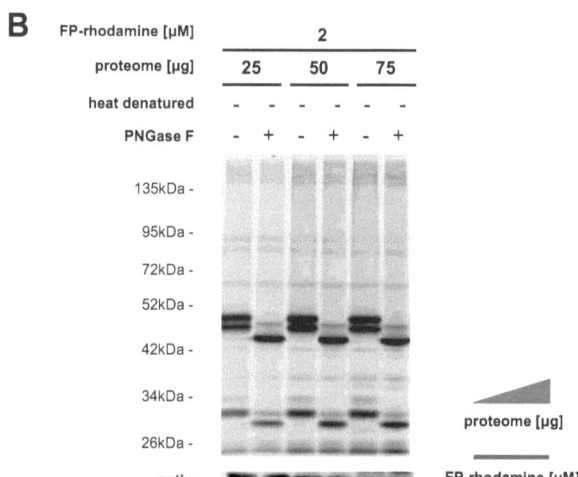

Figure 8B - SH activity profiles of CaLu-3 cell extracts (membrane fraction) at increasing proteome loading amounts. Increasing proteome loading amounts did not lead to the detection of additional SH activities.

Finally, SH activity profiles of CaLu-3 derived proteomes incubated at increasing FP-rhodamine concentrations with at the same time increasing proteome loading amounts are shown in Figure 8C. No significant increase in the number of detected SH activities was observed in this experiment. Based on the results presented above, it was decided to carry out future 1D-SDS-PAGE dependent activity-based experiments of membranous cell extracts at 2µM FP-rhodamine incubation concentration and use 25µg ABP labeled proteome as a loading amount.

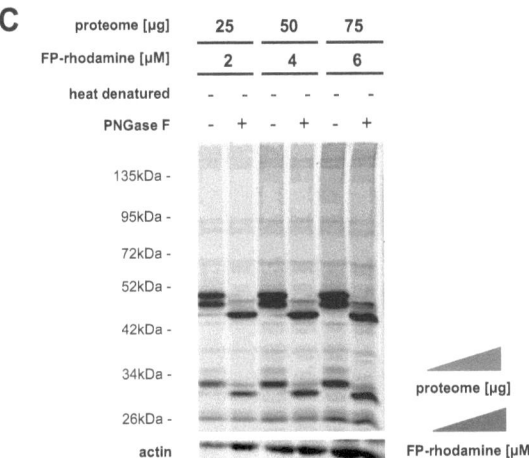

Figure 8C - SH activity profiles of CaLu-3 cell extracts (membrane fraction) at increasing FP-rhodamine labeling concentrations and increasing proteome loading amounts. The increase of proteome loading amounts combined with increasing FP-rhodamine concentrations did not lead to the detection of additional SH activities.

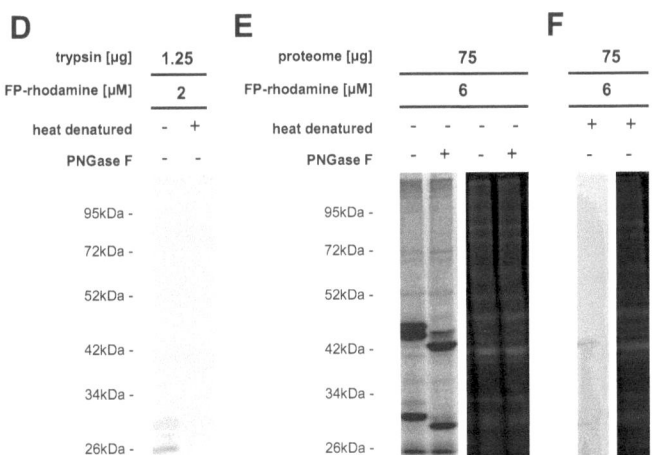

Figure 8D - Labeling efficiency assessment through parallel FP-rhodamine incubation of the SH trypsin (Promega). Figure 8E - SH activity profiles did not correspond with mere protein abundances (right side: Coomassie Blue (BioRad) staining of the same gel). **Figure 8F - Heat denatured proteomes labeled with FP-rhodamine served as negative controls** (right side: Coomassie Blue (BioRad) staining of the same gel).

In 1D-SDS-PAGE dependent experiments, FP-rhodamine labeling efficiencies were monitored through parallel incubation of the SH trypsin (Promega) and a heat denatured version thereof (see Figure 8D). Figure 8E shows that detected SH activities did not correspond to mere protein abundances. Heat denatured proteomes incubated with FP-rhodamine served as negative controls (see Figure 8F).

3.1.2 SH activity profiles of CaLu-3 cell extracts (soluble fraction)

Cell extracts were prepared as described in section 2.2. Labeling conditions as outlined in section 2.2 were employed to determine optimal labeling conditions of the soluble fraction of CaLu-3 derived cell extracts. Incubation of soluble cell extracts at standard labeling conditions (2μM FP-rhodamine, 25μg loading amount) resulted in the detection of approximately 10 distinct bands that correspond directly to enzymatic activities (Figure 9A). As illustrated in Figure 9A, the incubation of proteomes at increasing FP-rhodamine labeling concentrations compared to standard labeling conditions does not result in the detection of additional SH activities in case of CaLu-3 derived soluble cell extracts.

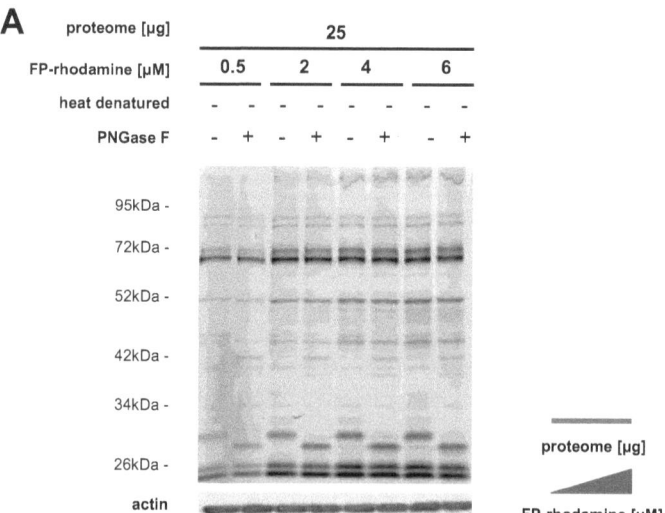

Figure 9A - SH activity profiles of CaLu-3 cell extracts (soluble fraction) at increasing FP-rhodamine labeling concentrations. Proteome incubation at 4μM and 6μM FP-rhodamine did not result in the detection of additional SH activities compared to labeling reactions at 2μM FP-rhodamine. PNGase F was used for protein deglycosylation. Actin served as a loading control.

Figure 9B shows detected SH activities at increasing proteome loading amounts incubated at 2µM FP-rhodamine. The results of this experiment revealed that a further increase in proteome loading amounts does not lead to the detection of additional SH activities. Taking into account that human samples are typically of limited quantity, it was decided to use a loading amount of 25µg per lane for soluble cell extracts derived from human biopsies for future experiments. Finally, SH activity profiles of CaLu-3 derived proteomes incubated at increasing FP-rhodamine concentrations with at the same time increasing proteome loading amounts, are shown in Figure 9C. No significant increase in the number of detected SH activities was observed in this experiment. Therefore, similar to the results presented in section 3.1.1, it was decided to carry out future 1D-SDS-PAGE dependent activity-based experiments of soluble fractions of cell extracts at 2µM FP-rhodamine labeling concentration and to use a loading amount of 25µg labeled proteome per lane. FP-rhodamine labeling efficiencies were monitored through parallel incubation of the SH trypsin (Promega) and a heat denatured version thereof (see Figure 9D). Figure 9E illustrates that detected SH activities of the soluble fraction did not correspond to mere protein abundances. Heat denatured proteomes incubated with FP-rhodamine served as negative controls (Figure 9F).

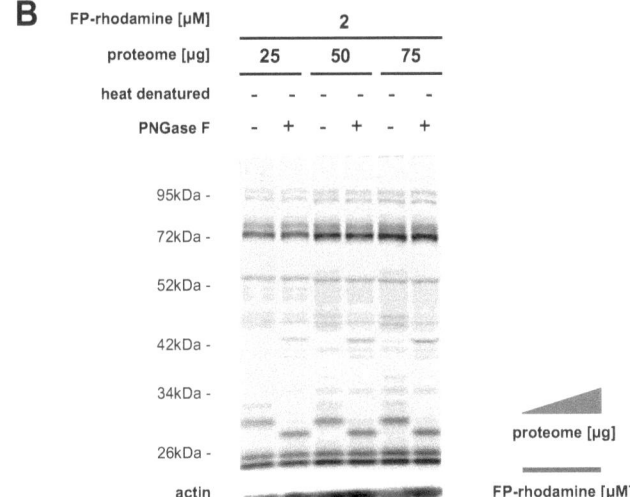

Figure 9B - SH activity profiles of CaLu-3 cell extracts (soluble fraction) at increasing proteome loading amounts. An increase in proteome loading amounts did not result in the detection of additional SH activities.

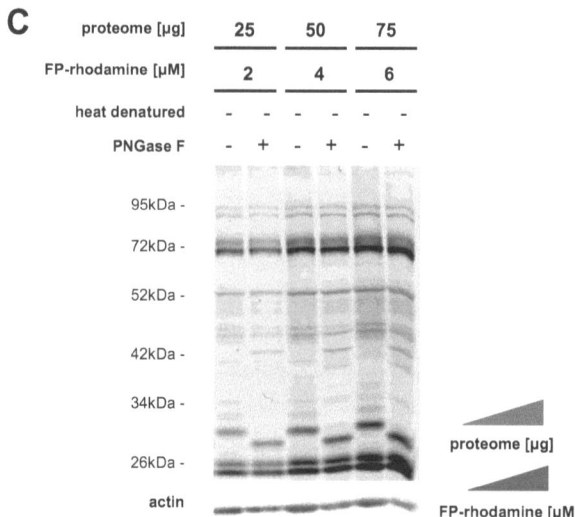

Figure 9C - SH activity profiles of CaLu-3 cell extracts (soluble fraction) at increasing FP-rhodamine labeling concentrations and increasing proteome loading amounts. Interestingly, the simultaneous increase of proteome loading amounts and activity-based probe labeling concentrations did not lead to a higher number of detected SH activities.

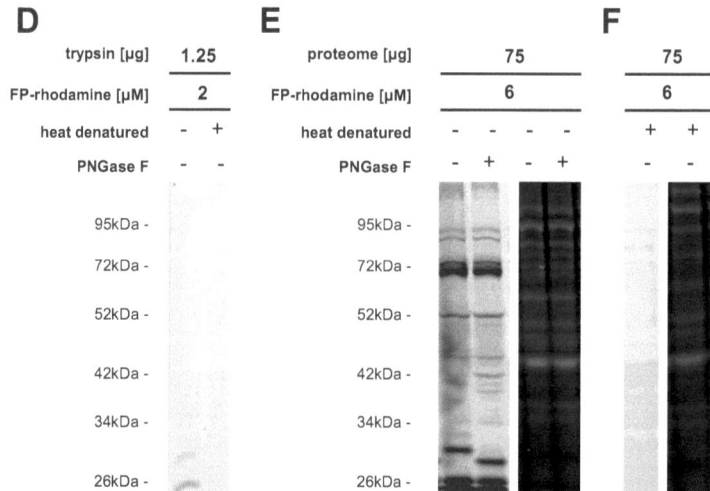

Figure 9D - Labeling efficiency assessment through parallel FP-rhodamine incubation of the SH trypsin (Promega). Figure 9E - SH activity profiles did not correspond with mere protein abundances in the soluble fraction (right side: Coomassie Blue (BioRad) staining of the same gel). **Figure 9F - Heat denatured proteomes labeled with FP-rhodamine served as negative controls** (right side: Coomassie Blue (BioRad) staining of the same gel).

3.1.3 SH activity profiles of human lung adenocarcinoma biopsies

After having established a 1D-SDS-PAGE dependent workflow for the determination of SH activity profiles in complex proteomes, a first cohort of human lung adenocarcinoma biopsies was investigated (SH activity profiles of 3 representative human lung adenocarcinoma proteomes are shown in Figure 10). During these experiments it was observed that the overall number of detected SH activities was reduced in human tumor proteomes compared to CaLu-3 derived proteomes (see section 3.1.1 and section 3.1.2).

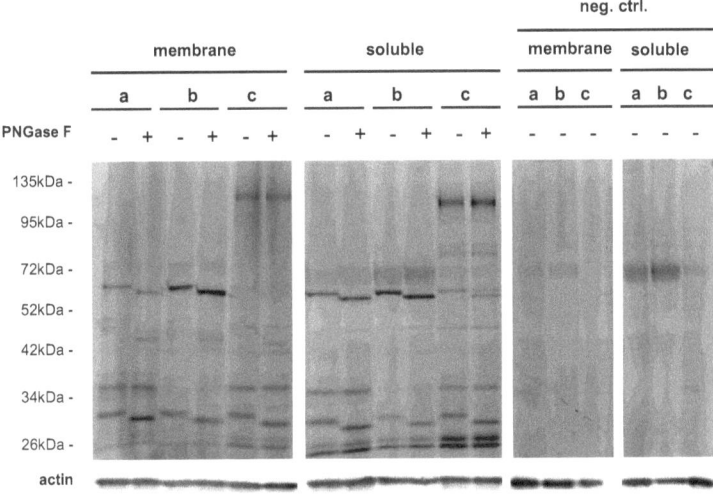

Figure 10 - SH activities in human lung adenocarcinoma. Cell extracts were separated into soluble and membrane fractions as described in section 2.2. Proteomes were incubated at standard labeling conditions (2µM FP-rhodamine, 1mg/ml proteome concentration) and 25µg of labeled proteomes were loaded per well on precast gels. The letters "a", "b" and "c" refer to biopsies of patients suffering from stage IB lung adenocarcinoma (6th TNM classification of malignant tumors, 2002). Tumor cell contents of investigated biopsies were 10%, 10% and 85%, respectively. Proteomes that served as negative controls were heat inactivated prior to activity-based probe incubation.

Based on the results presented above we hypothesized that the age of surgical specimens investigated within this experiment might have impaired proteome quality and consequently led to a reduced number of detectable enzymatic activities. However, a comparable number of SH activities was detected in proteomes derived from human lung adenocarcinoma biopsies as old as one month and specimens as old as seven years (stored at -80°C). Additionally, we found that the embedment of biopsies in an optimal cutting temperature (OCT) compound (Tissue Tek) as described in section 2.1 did also not impair SH activities. The exact reason for the decreased number of detected SH activities in human tumor proteomes compared to CaLu-3 derived proteomes remains therefore elusive, however, it can be speculated that abundant proteins derived from blood or other cell types present in human biopsies might have impaired the determination of protein concentration and consequently led to a reduced total amount of tumor cell specific SHs when standard labeling conditions were applied.

However, activity-based proteomics has already been introduced as a high-throughput platform for the investigation of SH activities in primary human specimens by Jessani et al. (Jessani et al., 2005). Thereby a quantitative, high-throughput and 1D-SDS-PAGE dependent analysis step is followed by a qualitative and highly sensitive LC/LC-MS/MS phase in which representative samples are analyzed (Jessani et al., 2005). As a consequence of this strategy, the qualitative analysis is dependent on the quantitative analysis and vice versa. The results presented above indicate that the 1D-SDS-PAGE dependent activity-based analysis is limited by a low dynamic range inherent to 1D-SDS-PAGE based experiments (see section 3.1). Therefore, this strategy does not benefit from highly sensitive instrumentation like LC/LC-MS/MS in case of the activity-based investigation of human lung adenocarcinoma biopsies. An alternative strategy with an optimized trade-off between throughput and sensitivity is therefore desirable for activity-based biomarker discovery studies in a clinical environment.

3.2 A mass spectrometry (MS) dependent platform for activity-based biomarker discovery

Here we describe the implementation of an advanced activity-based biomarker discovery platform that utilizes FP-1 (see section 2.4) as an activity-based probe to investigate SH activities in human tissues. The directed and solely MS dependent methodology employs *Progenesis LC-MS* for label-free quantification and allows simultaneous identification and relative quantification of active SHs in complex proteomes.

3.2.1 Considerations for a solely MS dependent activity-based biomarker discovery platform

Based on the results presented in section 3.1.3 it was decided to implement a solely LC/LC-MS/MS dependent activity-based biomarker discovery platform. However, an MS analysis time of 10hrs per LC/LC-MS/MS run (Prof. Cravatt, personal communication) renders this approach not feasible for clinical biomarker discovery where studies typically aim at analyzing a high number of samples and replicates. This is illustrated in the following example: the solely LC/LC-MS/MS dependent activity-based investigation of one patient, i.e. soluble and membranous cell extracts derived from one lung adenocarcinoma biopsy and one matching non-neoplastic lung tissue measured in triplicates with one negative control each requires a LC/LC-MS/MS analysis time of in total 160hrs.

Alternatively, the implementation of a LC-MS/MS dependent activity-based biomarker discovery platform with MS analysis times of 1.5hrs per LC-MS/MS run would significantly reduce the total analysis time required for the activity-based investigation of one patient, i.e. one lung adenocarcinoma biopsy and one matching non-neoplastic lung tissue. Additionally, the direct analysis of whole cell extracts was considered to be advantageous for biomarker discovery studies for two reasons: first, a reduction in the number of investigated cell extract fractions corresponds directly to a reduction in MS analysis time. Second, incomplete separation of whole cell lysates into soluble and membranous fractions by ultracentrifugation as illustrated in Figure

10 might impair quantification. In summary, for the complete qualitative and quantitative activity-based investigation of biopsies derived from one patient as described above, a solely LC-MS/MS dependent workflow would reduce MS machines times to 12hrs per patient.

3.2.2 Activity-based probes: FP-biotin and FP-1

The fluorophosphate derivative FP-1 (see section 2.4 and Figure 7) served as an activity-based probe in this study. FP-1 has been developed and successfully employed in several studies for the inhibition and isolation of active SHs (Higson et al., 1999; Nomura et al., 2005; Quistad and Casida, 2004). As shown in Figure 6 and Figure 7, the activity-based probe employed in this study (FP-1) and the activity-based probe introduced by Prof. Cravatt and coworkers (FP-biotin) differ in structure (Jessani et al., 2005).

In more detail, FP-1 represents a fluorophosphate derivative and FP-biotin represents a fluorophosphonate derivative. Both, fluorophosphate and fluorophosphonate derivatives inhibit the majority of SHs irreversibly in a covalent manner, whereas other hydrolases or inactive SHs remain unlabeled (Liu et al., 1999; Creighton, 1993; Walsh, 1979). One important difference between the two probes is that the distance between the fluorophosphate / fluorophosphonate groups and the biotin moiety, the so-called linker region, differs in length (see Figure 6 and Figure 7). Theoretically, the shorter linker region could be insufficient to span the distance between the active site serine of certain SHs and the protein exterior leaving the biotin within the protein framework.

To assess qualitative differences of the two activity-based probes, two independent batches of CaLu-3 derived cell extracts consisting of 3 biochemical replicates each were incubated with FP-1 and FP-biotin, respectively. Activity-based probe targeted SHs were enriched as described in section 2.5 and resulting tryptic peptide mixtures were analyzed on an LC-MS/MS system as described in section 2.6. Analysis of FP-1 labeled CaLu-3 proteomes resulted in the identification of 36 SHs (Mascot ion score > 25) and analysis of FP-biotin labeled CaLu-3 proteomes resulted in the identification of 38 SHs (Mascot ion score > 25). SHs identified in negative controls were excluded from analysis. Importantly, almost 70% of detected

SHs got indentified in FP-biotin labeled proteomes and in FP-1 labeled proteomes, indicating that both structures serve as activity-based probes with a broad reactivity against active SHs. At the same time these results demonstrate that the investigation of SH activities in complex proteomes with LC-MS/MS leads to a higher number of identified SH activities compared to analyses with 1D-SDS-PAGE (see section 3.1.1 and section 3.1.2). Conclusively, these results indicate that a solely LC-MS/MS dependent approach is better suited to be employed in clinical biomarker discovery studies compared to strategies where activity-profiles are determined by separated quantitative and qualitative analysis in which quantitative analysis depends on qualitative analysis and vice versa.

Based on reports describing the stability of FP-biotin, it was important to exclude the possibility that a degradation of FP-1 impairs activity assays conducted within this study (Schopfer et al., 2005). It was assumed that the most important chemical reaction that might render FP-1 inactive is the hydrolysis of the P-F bond (Schopfer et al., 2005). Therefore, the stability of the P-F bond of FP-1 was monitored by ^{19}F-NMR. These experiments proved that FP-1 remained stable upon reconstitution (5mM in DMSO) during the whole course of activity-based experiments conducted within this study (see Figure 11).

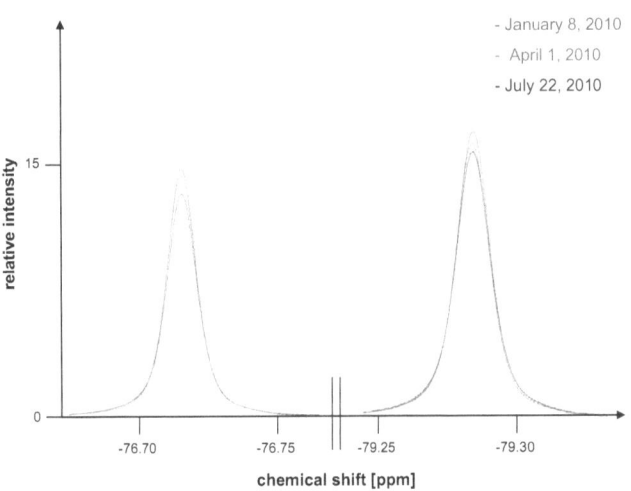

Figure 11 - Stability of the fluorophosphate derivative FP-1. ^{19}F-NMR revealed that the activity-based probe FP-1 remained stable upon reconstitution during the whole course of experiments.

In more detail, upon reconstitution in DMSO at 5mM final concentration, degradation of the P-F bond was monitored by ^{19}F-NMR measurements on January 8, 2010, April 1, 2010 and July 22, 2010, respectively. The integrated areas under the duplets from the measurements on April 1, 2010 and July 22, 2010 standardized by the area calculated by the measurement on January 8, 2010 resulted in a ratio of approximately 1. Taking small measurement errors into account it can be concluded that the P-F bond of the compound FP-1 remained stable upon reconstitution during the whole course of experiments.

3.2.3 Directed identification of active SHs in complex proteomes

In this study streptavidin coated agarose beads (Pierce) were utilized to enrich FP-1 targeted SHs out of complex proteomes before subsequent MS analysis (see section 2.5). This purification step led to a significant reduction in proteome complexity. However, after on-bead digestion we observed high amounts of tryptic peptides matching to streptavidin in the respective samples. Therefore, inclusion lists for direct sequencing and exclusion lists containing peptide sequences matching to streptavidin (Pierce) and trypsin (Promega) were generated as described in section 2.6. This strategy resulted in the identification of more than 80 distinct SHs in FP-1 labeled proteomes derived from lung adenocarcinoma biopsies and the human lung adenocarcinoma cell line CaLu-3 (for a complete list of identified enzymes see Supplemental Table S2). On average, 242 +/- 85 peptides (mean number of identified peptides +/- standard deviation) matching the criteria for quantification (see section 2.7) were detected per sample and led to the identification of approximately 40 SHs. The majority of identified SHs were esterases (EC 3.1.-.-) and proteases (EC 3.4.-.-) as determined by literature search and sequence comparison between identified proteins and conserved protein domains (Marchler-Bauer et al., 2009; Marchler-Bauer et al., 2007; Marchler-Bauer et al., 2002). Interestingly, several threonine proteases constituting the proteasome were identified, indicating that this class of enzymes is also susceptible to FP-1 (Jessani et al., 2005).7

3.2.4 Label-free quantification with *Progenesis LC-MS*

Validity of label-free quantification was confirmed by spiking defined amounts of the FP-1 labeled serine hydrolase murine Plasmin (UniProtKB / Swiss-Prot ID: P20918, Haematologic Technologies, Inc.) into FP-1 labeled lung tumor proteomes and subsequent analysis as described in section 2.5, section 2.6 and section 2.7. The top three intense peptide ions matching to murine Plasmin exhibited a good correlation between relative intensity and protein amount (see Figure 12). Similar results were obtained for less intense peptide ions detected in the same experiment (Supplemental Figure S2). In conclusion, label-free quantification of LC-MS data with *Progenesis LC-MS* is valid for this experimental setup for peptides exhibiting relative intensities ranging from 1E+02 to 2E+06.

However, during these experiments moderate variation of relative intensities of biochemical and even technical replicates with variation increasing inversely proportional to protein concentration have been observed. In subsequent experiments human tumor proteomes were therefore analyzed in sextuplicates (each biochemical triplicate measured in technical duplicates, see section 2.5) and the threshold for minimal relative peptide intensity was set to 2.5E+02. For all experiments conducted in this study, the vast majority of peptides used for quantification exhibited relative intensities that lied within the range of 2.5E+02 to 2E+06.

3.2.5 Repeatability, reproducibility and robustness

Repeatability was assessed through subsequent analysis of 15 biochemical replicates of an FP-1 labeled human lung tumor proteome using inclusion / exclusion lists as described in section 2.6. Data analysis with Scaffold version 2 (Proteome Software) identified 32 SHs with a similar number of peptides matching to a given protein in all 15 runs (peptide probability > 90%; median standard deviation for the number of identified peptides per protein: 20%). These results proved the LC-MS/MS system stable for a time frame required for complete proteomic analysis of one patient.

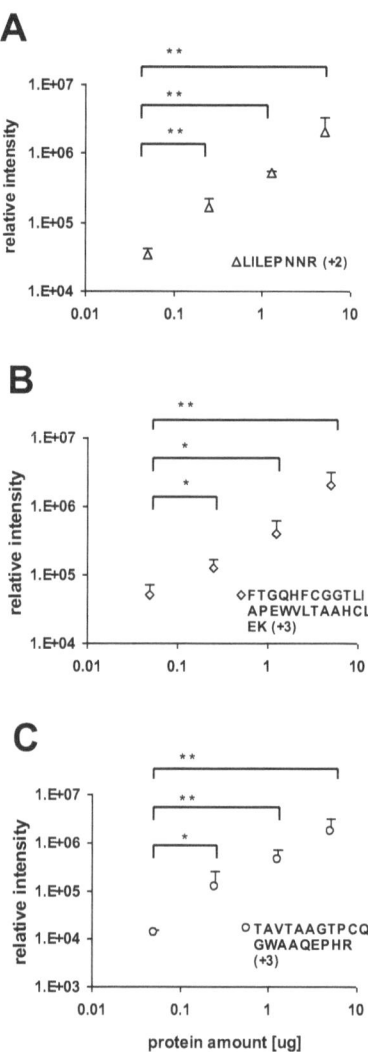

Figure 12 - Relative quantification of the top three intense peptide ions matching to the FP-1 labeled SH murine Plasmin in a complex tumor proteome. 12A) The top intense peptide ion exhibits a good correlation between relative intensity and protein amount. 12B - 12C) Similar results were obtained for the second and third intense peptide ion. Samples were measured in biochemical triplicates (*: $p<0.05$; **: $p<0.005$; unpaired, two sided Student`s t-test with log2 transformed values).

In order to assess reproducibility, tryptic peptide mixtures derived from three patients (patients #33, #48 and #56) after six months time were reanalyzed and data analysis performed as described in section 2.7. Again, inclusion / exclusion lists as described in section 2.6 were employed. In this experiment 90%, 86% and 83% of serine hydrolases initially detected in patients #33, #45 and #56, respectively, were repeatedly identified (FDR < 5% on protein level, FDR < 1% on peptide level). The average difference in fold-change of SH activities for patient #33 between the two measurements was 0.1 (log2 transformed value) with a standard deviation of 0.4 (log2 transformed value, see Figure 13). For patient #45, the average SH activity fold-change difference between the two experiments was -0.4 (log2 transformed value) with a standard deviation of 0.3 (log2 transformed value, see Figure 14). Finally, the average difference in fold change of SH activities for patient #56 was 0.2 (log2 transformed value) with a standard deviation of 0.3 (log2 transformed value, see Figure 15). In summary, the presented results provide evidence that the implemented workflow represents a repeatable and reproducible platform for the investigation of SH activities in complex proteomes.

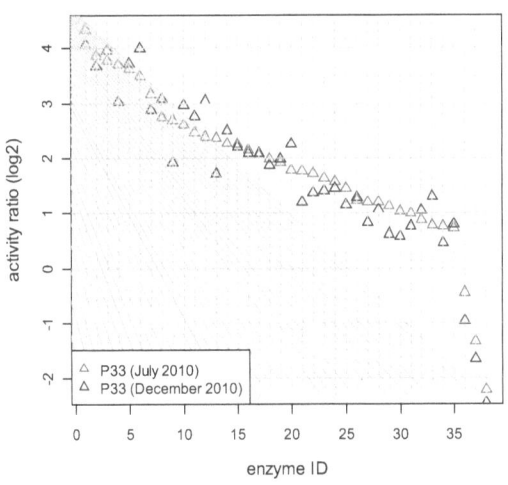

Figure 13 - Reproducibility of label-free LC-MS/MS dependent activity-based proteomics (patient #33). Tryptic peptide mixtures derived from patient #33 were reanalyzed after a time span of six months. The SH activity ratios detected in the second measurement were in good agreement with the ratios detected in the initial experiment.

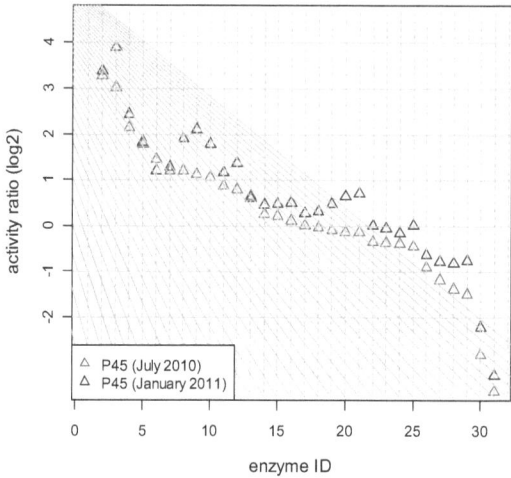

Figure 14 - Reproducibility of label-free LC-MS/MS dependent activity-based proteomics (patient #45). Tryptic peptide mixtures derived from patient #45 were reanalyzed after a time span of six months. The SH activity ratios detected in the second measurement were in good agreement with the ratios detected in the initial experiment.

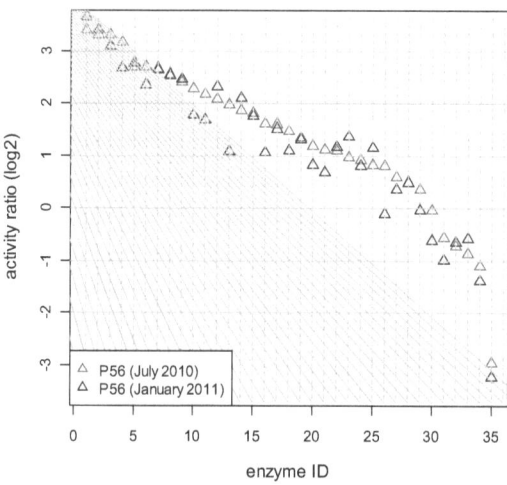

Figure 15 - Reproducibility of label-free LC-MS/MS dependent activity-based proteomics (patient #56). Tryptic peptide mixtures derived from patient #56 were reanalyzed after a time span of six months. The SH activity ratios detected in the second measurement were in good agreement with the ratios detected in the initial experiment.

Finally, MS-based analysis of 40 pairs of malignant and matching non-neoplastic tissues was conducted over a time frame of 8 months. Thereby it was observed that the number of SHs identified per sample remained stable (39 +/- 6.4, mean number of identified SHs +/- standard deviation). Furthermore, most SHs were repeatedly detected in the majority of samples (see Figure 16). These findings emphasize the robustness of the implemented activity-based biomarker discovery platform.

3.3 SH activity profiles in human lung adenocarcinoma

Forty pairs of human lung adenocarcinoma biopsies and matching non-neoplastic lung tissues were analyzed using the activity-based workflow described above. Two cases were excluded from analysis because of incorrect alignment of LC-MS features by *Progenesis LC-MS* ("patient 47" and "patient 22"). Data were analyzed with an unsupervised hierarchical clustering algorithm (*R* version 2.8.1, *R* packages *gplots*, *gdata* and *gtools* using default parameters, for *R* code see section 2.8). Only enzymes detected in at least 29 out of 38 patients were included into analysis resulting in a total number of 33 enzymes (see Figure 16). Classifications of patients obtained from the clustering algorithm were compared with stage, lymph node status and tumor grade.

"Stage" refers to the grouping of patients according to tumor size, lymph node status and presence of metastases at diagnosis (Dusmet, 2008). Increasing values of stage (I - IV) are directly proportional to worse prognosis (Minna, 2005). Patients with identical stage did not group together according to SH activity profiles (see Figure 16).

"Lymph node status" describes the absence (N0) or presence (N1, N2 or N3) of lymph node metastases at diagnosis (Mountain and Dresler, 1997). Increasing numbers indicate increasing distance of abnormal lymph nodes to the primary tumor and correspond to worse prognosis (Mountain and Dresler, 1997). Patients with identical lymph node status were not stratified into cohorts according to SH activity profiles (see Figure 16).

The process of "grading" refers to histological tumor classification according to the degree of differentiation (Suster, 2007). Patients with poorly differentiated (G3) tumors at diagnosis have worse prognosis and suffer from local recurrence more frequently compared to patients with well differentiated or moderately differentiated tumors (G1 and G2, respectively) (Barletta et al., 2010). Although patients with equivalent tumor grades clustered together in subgroups according to SH activity profiles, no convincing stratification of patients according to tumor grade was observed (see Figure 16).

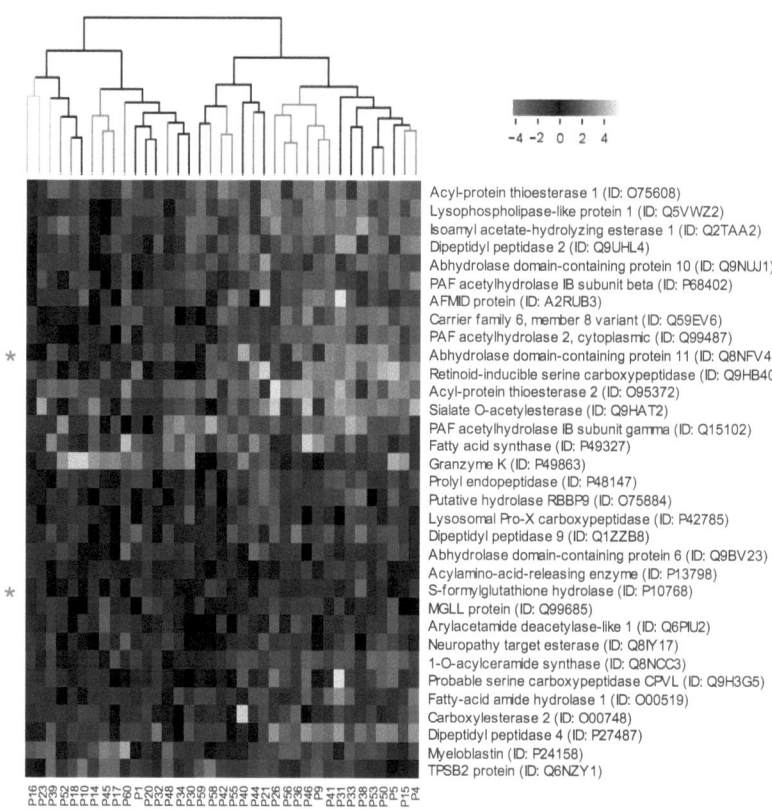

Figure 16 - Serine hydrolase activity profiles in human lung adenocarcinoma. Color key heatmap: green = elevated activity in malignant versus matching non-neoplastic tissue; red = decreased activity in malignant versus matching non-neoplastic tissue; black = no activity difference; blue = enzyme not detected (all values log2 transformed). Color key dendrogram: red = G3; grey = G2; black = G1 (see section 3.3). Red stars indicate SH activities of predictive biomarker candidates as described in section 3.4 and section 4.2.3. One patient ("patient 35") was excluded from illustration, because grading was not available (see Table 1 and Supplemental Table S1). All IDs refer to the UniProtKB / Swiss-Prot database.

3.4 Biomarker identification

By making use of logistic regression (R version 2.8.1, for R code see section 2.8) the activity of S-formylglutathione hydrolase (FGH), also known as Esterase D (ESD) (UniProtKB / Swiss-Prot ID: P10768) was found to statistically significantly predict the presence of poorly differentiated tumors (G3) compared to well differentiated (G1) or moderately differentiated tumors (G2) (n(G3) = 14, n(G2, G1) = 15, p<0.05). The activity of ESD discriminated poorly differentiated (G3) tumors (cut-off value: P = 0.55, see Figure 17) with a misclassification rate of 31%, a specificity of 73% and a sensitivity of 64% as determined by leave-one-out cross-validation and ROC analysis (AUC: 0.76, see Figure 19).

The potential of SH activities to predict the development of distant metastases (DMs) was also investigated. Participating individuals underwent complete surgical resection for lung adenocarcinoma and did either not develop DMs and received no adjuvant treatment or developed DMs and received adjuvant therapy or developed DMs and received no adjuvant treatment. None of the patients received neoadjuvant therapy. Minimum follow-up time for patients that did not develop DMs was 20 months. Statistical analysis with logistic regression revealed that the activity of the previously uncharacterized protein Abhydrolase domain-containing protein 11, Isoform 1 (ABHD11, UniProtKB / Swiss-Prot ID: Q8NFV4) statistically significantly predicts the development of DMs of patients with lymph node metastases (N1, N2) at diagnosis before undergoing radical surgery (n(DM) = 7, n(control) = 12, p<0.05). The activity of ABHD11 discriminated patients (cut-off value: P = 0.4, see Figure 18) that developed DMs and exhibited lymph node metastases at diagnosis before undergoing radical surgery with a misclassification rate of 21%, a specificity of 83% and a sensitivity of 71% as determined by leave-one-out cross-validation ROC analysis (AUC: 0.77, see Figure 20).

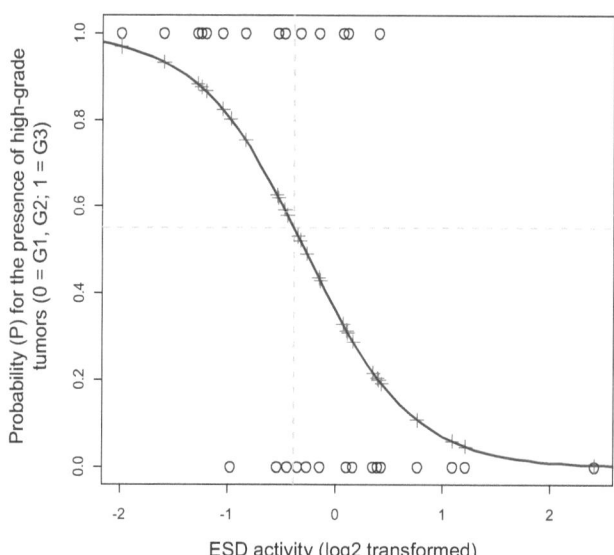

Figure 17 - ESD activities predicting the presence of high-grade tumors (G3) at a cut-off value of P = 0.55. Optimal trade-off between sensitivity and specificity was calculated by ROC analysis (see Figure 19). As indicated above, a cut-off-value of P = 0.55 corresponds to a 0.38-fold (log2 transformed) decreased ESD activity in malignant compared to matching non-neoplastic tissue. This value serves as a cut-off point in this model to discriminate patients that are more likely to suffer from high-grade tumors (G3).

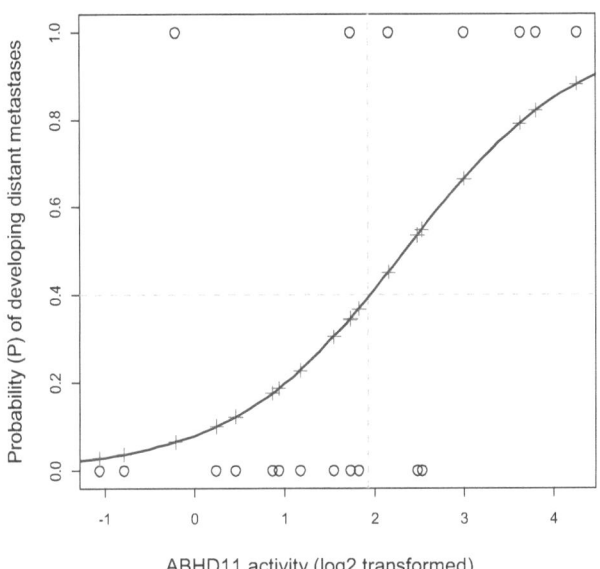

Figure 18 - ABHD11 activities predicting the development of distant metastases in patients with completely resected lung adenocarcinoma at a cut-off value of P = 0.4. Optimal trade-off between sensitivity and specificity was calculated by ROC analysis (see Figure 20). As indicated above, a cut-off-value of P = 0.4 corresponds to a 1.93-fold (log2 transformed) increased ABHD11 activity in malignant compared to matching non-neoplastic tissue. This value serves as a cut-off point in this model to discriminate patients that have higher chances of developing distant metastases.

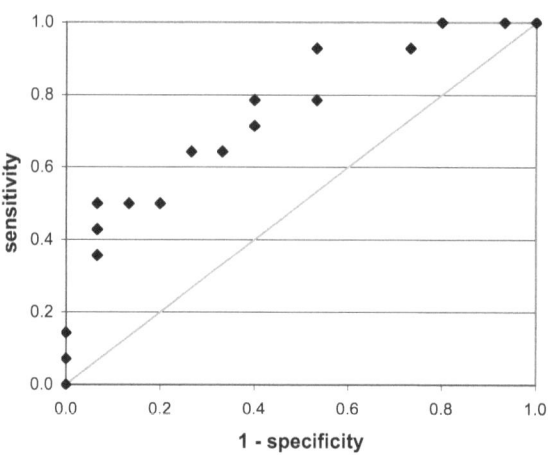

Figure 19 - Sensitivity and specificity of ESD activities predicting the presence of high-grade (G3) tumors. AUC: 0.76 as calculated using the trapezoidal rule.

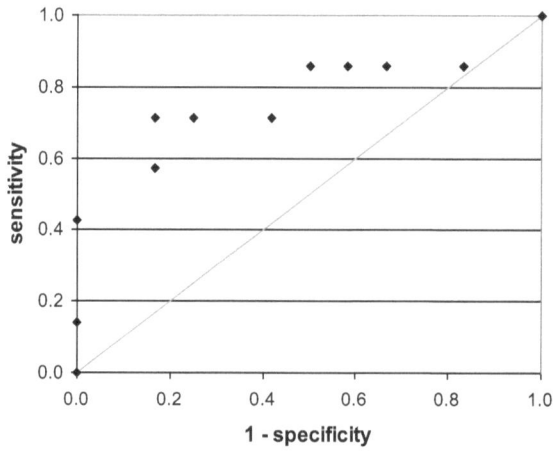

Figure 20 - Sensitivity and specificity of ABHD11 activities predicting the development of distant metastases in patients with completely resected lung adenocarcinoma. AUC: 0.77 as calculated using the trapezoidal rule.

3.5 Comparison of activity- and transcript-levels of ESD and ABHD11

Thousands of gene expression profiling experiments are conducted every year and many of the data are made publicly available by the scientific community (Kilpinen et al., 2008). However, due to a high number of different technologies used for conducting gene expression profiling experiments, it is often challenging to compare results between different studies (Kilpinen et al., 2008). This shortcoming was addressed by Kilpinen et al. through the implementation of an algorithm that normalizes data arising from different Affymetrix microarray generations (Kilpinen et al., 2008).

Making use of this algorithm that has been implemented in a publicly available human gene expression database (*www.genesapiens.org*), no differences of *ESD* transcript levels in normal tissues of the respiratory system compared to malignant lung tissues were found (Kilpinen et al., 2008). Each dot in Figure 21 represents the relative expression level of *ESD* in one sample (Kilpinen et al., 2008). None of the samples presented in Figure 21 exhibits a significant higher than average expression or represents an outlier expression profile (Kilpinen et al., 2008). However, slightly elevated *ABHD11* expression levels were found in healthy tissues of the respiratory system compared to malignant lung tissues (see Figure 22) (Kilpinen et al., 2008).

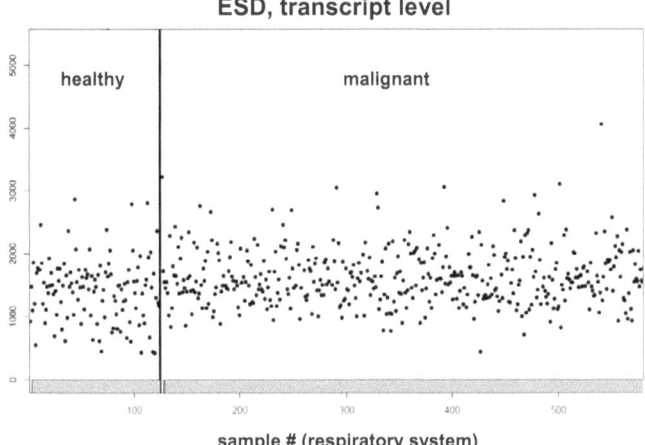

Figure 21 - *ESD* transcript levels in healthy tissues of the respiratory system compared to malignant lung tissues do not show significant differences. The plot was downloaded from *www.genesapiens.org* on September 15, 2010.

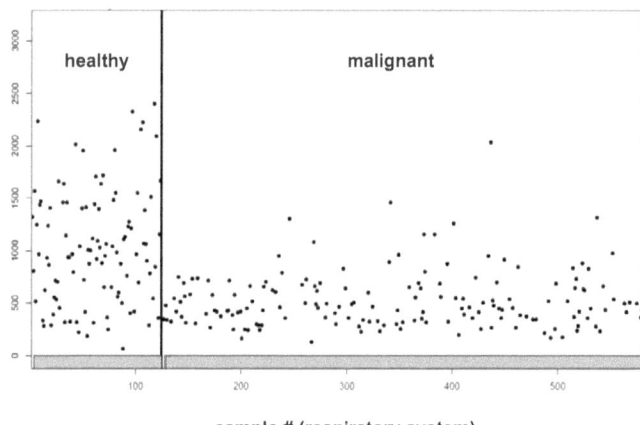

sample # (respiratory system)

Figure 22 - *ABHD11* transcript levels in healthy tissues of the respiratory system compared to malignant lung tissues do not show significant differences. A slightly higher level of *ABHD11* expression was observed in healthy tissues of the respiratory system. The plot was downloaded from *www.genesapiens.org* on September 15, 2010.

4. Discussion

In this study an advanced activity-based proteomics platform for clinical biomarker discovery has been presented. Forty pairs of malignant and matching non-neoplastic lung tissues were investigated for serine hydrolase (SH) activities and two predictive biomarker candidates that have previously not been associated with NSCLC, namely Esterase D and Isoform 1 of Abhydrolase domain-containing protein 11, have been identified.

4.1 Activity-based proteomic strategies for the investigation of SH activities in complex proteomes

An activity-based proteomics platform for the investigation of SH activities in human specimens has previously been implemented by Prof. Cravatt and colleagues (Jessani et al., 2005). The strategy pursued in this approach is described in the following: first, a high-throughput and 1D-SDS-PAGE dependent analysis step is employed for quantitative analysis (Jessani et al., 2005). Second, a highly sensitive MS-based (LC/LC-MS/MS) analysis step where representative samples are analyzed is used for protein identification (Jessani et al., 2005).

The first goal of this study was the implementation of the 1D-SDS-PAGE dependent activity-based analysis platform as described by Jessani et al. for the investigation of SH activities in human specimens (see section 3.1) (Jessani et al., 2005). Initial experiments investigating proteomes derived from the human lung adenocarcinoma cell line CaLu-3 resulted in SH activity profiles that were qualitatively, i.e. the number of detected SH activities, comparable to SH activity profiles derived from human breast cancer tissues as reported by Jessani et al. (Jessani et al., 2005). Experiments aiming at increasing the number of detectable SH activities in CaLu-3 derived proteomes through variation of activity-based probe concentration, proteome loading amount or both were conducted (see section 3.1.1 and section 3.1.2). From the results obtained in these experiments it was concluded that an increase in activity-based probe concentration, proteome loading amount or both does not lead to a substantial improvement in sensitivity of 1D-SDS-PAGE dependent activity-based proteomics investigating CaLu-3 derived proteomes.

Nonetheless, by making use of the implemented workflow, proteomes derived from human lung adenocarcinoma biopsies were investigated (see section 3.1.3). Interestingly, the number of SH activities detected within these experiments was significantly reduced compared to SH activity profiles detected in human breast cancer specimens as reported by Jessani et al. (Jessani et al., 2005). It can be speculated that differences in tissue specific SH expression in human lung / lung cancer and human breast / breast cancer tissues might be the reason for this observation.

The primary aim of this study was the identification of SH activities as potential biomarker candidates in human lung adenocarcinoma. Based on the results discussed above, it was anticipated that 1D-SDS-PAGE dependent activity-based proteomics is a suboptimal methodology to assess the predictive potential of SH activities in human lung adenocarcinoma since the depth of the analysis is restricted by the dynamic range inherent to 1D-SDS-PAGE. Therefore, this approach does not benefit from highly sensitive instrumentation like LC/LC-MS/MS for qualitative analysis. Hence, an alternative strategy with an optimized trade-off between throughput and sensitivity is desirable for activity-based biomarker discovery studies in a clinical environment.

4.1.1 An advanced strategy for the identification of SH activities in complex proteomes

In order to establish an activity-based workflow feasible for biomarker discovery in a clinical setting, a solely LC-MS/MS dependent analysis platform based on the considerations outlined in section 3.2.1 has been established. In this study, FP-1 (see section 2.4 and Figure 7) served as an activity-based probe. FP-1 has initially been developed and subsequently employed for the inhibition and isolation of active SHs (Higson et al., 1999; Nomura et al., 2005; Quistad and Casida, 2004). Direct comparison of the number of SHs targeted by FP-1 and FP-biotin, another activity-based probe that has successfully been employed to study activity states of SHs, revealed a 70% overlap in enzymes targeted by both activity-based probes (see section 3.2.2) (Jessani et al., 2005). These results indicate that FP-1 indeed serves as a valid activity-based probe for the investigation of SH activity states in complex proteomes.

The implemented, solely LC-MS/MS dependent strategy consists of a labeling step, a purification step and a mass spectrometry based protein identification step. During the labeling step, proteomes are incubated with FP-1, in the purification step FP-1 targeted enzymes are enriched by making use of streptavidin coated agarose beads followed by on-bead digestion. In the last step, enzymes are identified and relatively quantified using mass spectrometry.

Although the purification step led to a significant reduction in proteome complexity, high amounts of tryptic peptides matching to streptavidin were observed in analyzed samples. Since shotgun proteomics is based on simple heuristics, i.e. tryptic peptides are chosen for analysis based on signal intensities in the survey scan, high levels of peptides matching to streptavidin impair the detection of peptides matching to low abundant SHs (Domon and Aebersold, 2010). Taking that into consideration, a directed strategy was pursued where a predetermined set of peptide ions are detected and selected for fragmentation (peptides matching to SHs) or excluded from fragmentation (peptides matching to streptavidin or trypsin, see section 3.2.3). Retention times (R_T) and mass-to-charge (m/z) ratios, two parameters that are valid identifiers for any peptide of interest, are summarized in so-called inclusion and exclusion lists, respectively. Inclusion and exclusion lists were generated from data derived from initial experiments in DDA mode and additional m/z values of peptides matching to SHs were added to these lists by making use of the publicly available database *PeptideAtlas* (Desiere et al., 2005).

Overall, more than 80 distinct SHs in FP-1 labeled proteomes derived from human lung adenocarcinoma biopsies and the human lung adenocarcinoma cell line CaLu-3 were identified by employing this strategy (for a comprehensive list of identified SHs see Supplemental Table S2). It has been reported that SHs comprise about 1% of the human proteome which translates into 210+ SHs (Simon and

Cravatt, 2010). Taking into account that this study focused on the investigation of a specific tissue, i.e. human lung / lung cancer biopsies, the repertoire of detected enzymes represents a substantial fraction of SHs present in the human proteome. In this context, it is important to mention that the use of inclusion lists for directed sequencing is of special importance considering that inclusion lists can be expanded upon the availability of additional peptide sequences matching to SHs.

4.1.2 Label-free quantification

Validity of label-free quantification was confirmed by spiking defined amounts of the FP-1 labeled SH murine Plasmin (see section 3.2.4) in a complex lung tumor proteome and subsequent analysis as described in section 2.5, section 2.6 and section 2.7.

The three peptide ions matching to murine Plasmin that exhibited the highest relative intensities as calculated by *Progenesis LC-MS* showed a good correlation between relative intensity and protein amount (see Figure 12). Equivalent results were obtained for peptide ions detected in the same experiment that exhibited significantly lower relative intensities (see Supplemental Figure S2). Although these results provide evidence that relative quantification with *Progenesis LC-MS* is valid for the experimental workflow implemented in this study, we want to point out that this strategy is limited by a dynamic range of approximately 2 to 3 orders of magnitude and moderate variation of relative intensities from biochemical and even technical replicates with variation increasing inversely proportional to protein concentration has been observed.

In future activity-based proteomics studies an increase in dynamic range might be accomplished by employing one of the following strategies: first, when sequence information of all enzymes of potential interest is available, quantitative selected reaction monitoring (SRM) measurements can be performed after SRM assay development. A dynamic range of up to five orders of magnitude has been reported for SRM-based experiments (Lange et al., 2008). However, so far not all serine hydrolases have been annotated as such. Therefore, we anticipate that SRM measurements will be best suited for activity-based biomarker validation studies

(Simon and Cravatt, 2010). Second, a technically challenging though elegant solution for activity-based proteomics involves the incubation of proteomes or even intact cells with a cell membrane permeable, alkyne containing activity-based probe (Weerapana et al., 2007). After ABP incubation, targeted enzymes are linked with a tobacco etch virus (TEV) biotin conjugate through a chemical reaction known as click-chemistry (Weerapana et al., 2007). The term click chemistry refers to a reaction where a copper-catalyzed analog of Huisgen's triazole synthesis covalently couples an alkyne to an azide (Weerapana et al., 2007). After enrichment of targeted enzymes by making use of avidin or streptavidin coated agarose beads, enzymes are recovered through incubation with TEV-protease (Weerapana et al., 2007). Since the exact cleavage site of the TEV sequence containing ABP construct is known, the complete primary sequence of the targeted enzyme including the active site peptide can be used for protein identification with MS (Weerapana et al., 2007). Additionally, in analogy to the strategy employed for precise relative quantification with isotope-coded affinity tags (ICAT), we anticipate that heavy and light versions of isotopically labeled and alkyne containing ABPs will allow accurate relative quantification between two samples, i.e. tumor and matching non-neoplastic tissue (Gygi et al., 1999a).

Although the implemented workflow described in this study exhibits moderate dynamic range and variations of relative intensities of low-intense peptide ions have been observed in repetitive measurements, it is important to mention that label-free quantification represents a experimentally feasible and cost-effective way to relatively compare SH activities between tumor and matching non-neoplastic tissues.

4.1.3 Repeatability, reproducibility and robustness

Repeatability of the implemented methodology was assessed through subsequent analysis of 15 biochemical replicates of an FP-1 labeled human lung tumor proteome. Thirty-two SHs with a similar number of peptides matching to a given SH were detected in all 15 runs (see section 3.2.5). The results of this experiment provide evidence that the LC-MS/MS system remains stable for the time frame required for complete activity-based proteomic analysis of one patient.

Reproducibility was assessed by reanalyzing tryptic peptide mixtures derived from three malignant and matching non-neoplastic lung tissues after a time span of six months. In all three experiments, more that 80% of SHs were repeatedly identified with low average differences in fold-change of SH activities between the two analyses (see section 3.2.5).

Finally, MS-based analysis of 40 pairs of malignant and matching non-neoplastic tissues was conducted over a time frame of 8 months. Thereby, the number of SHs identified per sample remained stable throughout the whole course of experiments (see section 3.2.5). Furthermore, most SHs were repeatedly detected in the majority of samples (see Figure 16). In summary, these results provide evidence for good repeatability and reproducibility of the implemented platform and emphasize the robustness of this activity-based biomarker discovery workflow.

However, it is important to mention that the purification protocol for FP-1 targeted SHs as outlined in section 2.5 consists of a large number of incubation and purification steps. Based on the results discussed above, we anticipate that the purification of FP-1 targeted SHs prior to mass spectrometric analysis is more prone to error than the mass spectrometric analysis itself. In our view, major sources of error that can influence quantitative and qualitative results are as follows: first, irregularities in cell lysate preparation. We recommend that the same person prepares all cell lysates investigated in a given activity-based proteomics study. Second, since native proteomes are investigated, we recommend to assess proteome qualities by 1D-SDS-PAGE and subsequent Commassie Blue staining before proteomes are incubated with the respective activity-based probe (see Supplemental Figure S1). Third, the stability of the activity-based probe, FP-1 incase of this study, should be monitored with nuclear magnetic resonance (NMR, see section 3.2.2). Fourth, activity-based probe labeling efficiencies should be assessed through parallel incubation of a model enzyme (the SH trypsin in case of this study) with the respective ABP and subsequent detection through Western Blot (see Supplemental Figure S1). Fifth, the efficiency of on-bead digestion of ABP targeted enzymes should ideally be monitored through parallel incubation of a model protein (bovine serum albumin (BSA), for example) with the respective digestion enzyme followed by analysis with 1D-SDS-PAGE and subsequent Commassie Blue staining.

4.1.4 Activity-based proteomics and biomarker discovery: technical aspects

Although the presented methodology does not reach the throughput achieved by activity-based workflows with separated quantitative (1D-SDS-PAGE) and qualitative (LC/LC-MS/MS) analysis, simultaneous identification and quantification of proteins is highly valuable in clinical biomarker discovery. The presented workflow is especially well suited to be employed in experimental setups where malignant and matching non-neoplastic tissues are investigated: first, taking the unique genetic background of every patient into account, activity signatures can be normalized on an individual basis (Marian, 2009). Second, advanced software solutions for label-free quantification of LC-MS data require significant computer processing power that increases directly proportional to the number of samples analyzed. Therefore, performing quantification for each patient individually is advantageous in biomarker discovery where studies typically aim at analyzing a high number of samples. Notably, in this study surgical specimens as old as seven years (stored at -80°C) have successfully been analyzed. This is an encouraging finding for future activity-based biomarker discovery studies where the availability of long patient follow-up times is desirable.

However, inherent to the presented methodology are the time-consuming LC-MS/MS analysis, especially when a high number of replicates is investigated and the high proteome consumption as a consequence thereof. This impedes the application of the presented workflow in a clinical laboratory working on a daily basis. However, we anticipate that previously described, easy-to-use microplate arrays may represent high-throughput tools for activity-based diagnostics (Sieber and Cravatt, 2006; Sieber et al., 2004).

General shortcomings of activity-based biomarker discovery are discussed in the following: first, due to the sensitivity of enzymatic activities to detergents, methods like laser capture microdissection (LCM) are not applicable for tumor tissue enrichment (see section Publications) (Collaud et al., 2010). Second, activity-based proteomics is limited to the investigation of one enzyme superfamily per analysis. Although a variety of activity-based probes have been developed, simultaneous analysis of several enzyme classes with activity-based probe cocktails has not been reported so far (Cravatt et al., 2008). Nonetheless, we believe that the unique properties of activity-based proteomics, i.e. direct enzymatic activity-readouts of virtually all kinds of tissues and body fluids are highly valuable in the search for novel disease biomarkers.

4.2 SH activities and human lung adenocarcinoma

4.2.1 Global SH activity profiles as potential biomarkers for human lung adenocarcinoma

Serine hydrolase activities derived from proteomic analysis of 38 out of 40 lung adenocarcinoma and matching non-neoplastic lung tissues were analyzed with an unsupervised hierarchical clustering algorithm (see section 3.3, for *R* code see section 2.8). Two cases were excluded from analysis because of incorrect alignment of LC-MS features by *Progenesis LC-MS* ("patient 47" and "patient 22"). For clustering analysis aiming at stratifying patients into cohorts according to tumor grade based on SH activity profiles, one patient ("patient 35") was excluded from analysis, because grading was not available (see Table 1 and Supplemental Table S1).

The hierarchical clustering analysis consisted of two steps: first, the clustering algorithm grouped patients together which exhibited similar SH activity patterns. Thereby it is important to note that this grouping is based on a mathematical description of similarity (Eisen et al., 1998). Since a large number of different mathematical measures of similarity exist, the findings shown in Figure 16 represent only one out of a number of possible results (Eisen et al., 1998). Second, the results as calculated by the clustering algorithm were manually linked with clinical characteristics of patients suffering from lung adenocarcinoma. Patients with identical stage or lymph node status were not stratified into cohorts according to SH activity profiles. Although patients with equivalent tumor grades clustered together in subgroups according to SH activity profiles, no convincing stratification of patients according to tumor grade was observed (see section 3.3). However, we want to point out that the limited sample size employed in this study might have impaired the correlation of global serine hydrolase activity profiles with clinical characteristics.

As illustrated in Figure 16, a number of SHs got repetitively identified during the course of this project. Therefore, before selected SHs are discussed in more detail, an overview of SHs detected in at least 29 out of 38 patients (see section 3.3) is given in Table 2. The content of Table 2 aims to provide general information on the biological function of the identified enzymes and, if available, their role in NSCLC.

Protein name	Function	UniProtKB / Swiss-Prot ID
1-O-acylceramide synthase	is a lysosomatic enzyme, also known as Lysosomal phospholipase A2, that catalyzes the transacylation of short chain ceramides (Shayman et al., 2011)	Q8NCC3
Abhydrolase domain-containing protein 10	see section 4.2.2	Q9NUJ1
Abhydrolase domain-containing protein 11	see section 4.2.3.2	Q8NFV4
Abhydrolase domain-containing protein 6	plays a role in neurotransmission and -inflammation and is differentially expressed on a transcript level among various cancer cell lines (Marrs et al., 2010; Li et al., 2009)	Q9BV23
Acylamino-acid-releasing enzyme	deletion of the genetic region encoding for Acylamino-acid-releasing enzyme has been associated with NSCLC (Mitta et al., 1996)	P13798
Acyl-protein thioesterase 1	see section 4.2.2	O75608
Acyl-protein thioesterase 2	shows 64% protein sequence similarity to Acyl-protein thioesterase 1 (see section 4.2.2), however, its biological function remains to be determined (Zeidman et al., 2009)	O95372
AFMID protein	a serine hydrolase involved in the degradation of tryptophane (Pabarcus and Casida, 2005)	A2RUB3
Arylacetamide deacetylase-like 1	see section 1.6	Q6PIU2
Carboxylesterase 2	Carboxylesterase 2, also known as Cocaine esterase according to the UniProtKB / Swiss-Prot database, is expressed at high levels in the colon (Jewell et al., 2007). Cocaine esterase catalyzes the hydrolysis of cocaine and bacterial Cocaine esterase has been shown to prevent cocaine-induced toxicity in rats (Liu et al., 2009a; Collins et al., 2009)	O00748
Dipeptidyl peptidase 2*	a serine protease that degrades oligopeptides and plays a role in the regulation of cell quiescence (Maes et al., 2005; Mele et al., 2009)	Q9UHL4

Protein name	Function	UniProtKB / Swiss-Prot ID
Dipeptidyl peptidase 4*	is a cell surface protease that has been suggested to act as a tumor suppressor (Wesley et al., 2004; Olsen and Wagtmann, 2002). Down-regulation of Dipeptidyl peptidase 4 has been shown to contribute to uncontrolled growth in NSCLC cell lines (Wesley et al., 2004)	P27487
Dipeptidyl peptidase 9*	shares sequence similarity with Dipeptidyl peptidase 4 and contains a serine protease motif (Olsen and Wagtmann, 2002). Dipeptidyl peptidase 9 has been shown to influence interactions between cells and the extracellular matrix (Yu et al., 2006)	Q1ZZB8
Fatty-acid amide hydrolase 1	represents a serine hydrolase with both, esterase and amidase activities (Patricelli and Cravatt, 1999). On a protein level, Fatty-acid amide hydrolase 1 expression is associated with prostate cancer severity (Thors et al., 2010)	O00519
Fatty acid synthase	see section 1.6	P49327
Granzyme K*	an enzyme that exhibits tryptase-like activities with similar substrate specificities (Cullen et al., 2010). Granzyme K has been shown to induce caspase independent cell death (Zhao et al., 2007)	P49863
Isoamyl acetate-hydrolyzing esterase 1	see section 4.2.2	Q2TAA2
Lysophospholipase-like protein 1	see section 4.2.2	Q5VWZ2
Carrier family 6, member 8 variant*	this lysosomal serine protease is encoded by the same gene locus as Lysosomal protective protein / Cathepsin A (UniProtKB / Swiss-Prot ID: P10619) as determined with the Ensembl Genome Browser (Hiraiwa, 1999). The biological relevance of Carrier family 6, member 8 variant remains to be determined	Q59EV6
Lysosomal Pro-X carboxypeptidase*	also referred to as Prolylcarboxypeptidase according to the UniProtKB / Swiss-Prot database, has been shown to regulate proliferation and autophagy in breast cancer cells (Duan et al., 2011)	P42785
MGLL protein	referred to as Monoacylglycerol lipase according to the UniProtKB / Swiss-Prot database, is expressed at elevated levels in aggressive cancer cells and influences migration and tumor growth through the regulation of a fatty acid network (Nomura et al., 2010b)	Q99685

Protein name	Function	UniProtKB / Swiss-Prot ID
Myeloblastin*	a serine protease that causes growth arrest and differentiation of leukemic cells (Bories et al., 1989)	P24158
Neuropathy target esterase	is a serine esterase that plays an important role in neural development and axonal maintenance (Chang et al., 2011; Glynn, 1999)	Q8IY17
PAF acetylhydrolase 2, cytoplasmic	Platelet-activating factor (PAF) is a lipid messenger that plays a role in a variety of physiological processes and cancer (Arai et al., 2002; Mills and Moolenaar, 2003). An acetyl group at a specific position on PAF is crucial for its biological activity (Arai et al., 2002). The removal of this acetyl group is catalyzed by PAF-acetylhydrolase (PAF-AH) (Arai et al., 2002). Several isoforms like PAF acetylhydrolase 2, cytoplasmic, PAF acetylhydrolase IB subunit beta or PAF acetylhydrolase IB subunit gamma have been identified (Chen, 2004)	Q99487
PAF acetylhydrolase IB subunit beta		P68402
PAF acetylhydrolase IB subunit gamma		Q15102
Probable serine carboxypeptidase CPVL*	has initially been found to be expressed in macrophages (Mahoney et al., 2001). The degradation of phagocytosed particles in the lysosome as well as a role in an inflammatory protease cascade are suggested biological functions for this enzyme (Mahoney et al., 2001)	Q9H3G5
Prolyl endopeptidase*	a serine protease that is involved in the modification of biologically active peptides (Larrinaga et al., 2010). A role of prolyl endopeptidase in neoplastic processes has been suggested (Larrinaga et al., 2010)	P48147
Putative hydrolase RBBP9	exerts elevated activities in pancreatic carcinomas (Shields et al., 2010). In malignant cells, the elevated activity of RBBP9 promotes anchorage-independent growth (Shields et al., 2010). In more detail, it has been shown that RBBP9 can overcome TGF-β-mediated antiproliferative signaling through reducing Smad2/3 phosphorylation (Shields et al., 2010)	O75884
Retinoid-inducible serine carboxypeptidase*	a lysosomal matrix protein whose biological function still needs to be determined (Kollmann et al., 2009; Kollmann et al., 2005)	Q9HB40
S-formylglutathione hydrolase	see section 4.2.3.1	P10768
Sialate O-acetylesterase	plays a role in bacterial degradation of mucin glycoproteins in the human colon (Corfield et al., 1992)	Q9HAT2

Protein name	Function	UniProtKB / Swiss-Prot ID
TPSB2 protein*	see section 4.2.2	Q6NZY1

Table 2 - Overview of SHs detected in human lung / lung cancer biopsies during the course of this study. Stars indicate serine peptidases / proteases as determined by literature search or sequence comparison between identified proteins and conserved protein domains (Marchler-Bauer et al., 2009; Marchler-Bauer et al., 2007; Marchler-Bauer et al., 2002).

4.2.2 SH activities associated with human lung adenocarcinoma

During data analysis it was observed that a number of SHs exhibited higher or lower activities in malignant versus matching non-neoplastic tissues in an almost exclusive manner (see Table 3 and Figure 16).

Protein name	UniProtKB / Swiss-Prot ID	Activity (malignant vs. normal)
TPSB2 protein	Q6NZY1	↓
Acyl-protein thioesterase 1	O75608	↑
Lysophospholipase-like protein 1	Q5VWZ2	↑
Isoamyl acetate-hydrolyzing esterase 1	Q2TAA2	↑
Abhydrolase domain-containing protein 10	Q9NUJ1	↑

Table 3 - SHs exhibiting almost exclusively higher or lower activities in malignant versus matching non-neoplastic tissues.

TPSB2 protein, the *TPSB2* gene product, is a serine hydrolase that shares high protein sequence similarity with Tryptase β-2, also a *TPSB2* gene product. TPSB2 protein (length: 282 amino acids, UniProtKB / Swiss-Prot ID: Q6NZY1) and Tryptase β-2 (length: 275 amino acids, UniProtKB / Swiss-Prot ID: P20231) share 274 identical and 1 similar amino acid positions as determined with the *clustalw* algorithm (sequences derived from *www.uniprot.org* on February 20, 2011) (Higgins and Sharp, 1988; Thompson et al., 1994). Although TPSB2 protein remains functionally uncharacterized, Tryptase β-2, an enzyme stored in the secretory granules of mast cells, acts as a growth factor for epithelial cells and airway smooth muscle cells (Payne and Kam, 2004; Ren et al., 1998). Furthermore, during inflammation, Tryptase β-2 stimulates the release of granulocyte chemoattractants and induces the expression of interleukin-1β which is thought to be involved in the recruitment of inflammatory cells to sites of mast cell activation (Payne and Kam, 2004; Ren et al., 1998). Mast cells are present in a variety of tumors and low mast cell densities found in the peritumoral zone of resected NSCLC specimens have been associated with poor prognosis (Carlini et al., 2010; Khazaie et al., 2011). TPSB2 protein exhibited lower activities in the vast majority of investigated lung adenocarcinoma biopsies, whereas *TPSB2* transcript levels did not show any differences in lung cancer biopsies versus normal tissues of the respiratory system as determined by the publicly accessible human gene expression database *www.genesapiens.org* (see Supplemental Figure S3) (Kilpinen et al., 2008). This observation indicates that differences on a transcript level do not necessarily correspond with changes in enzymatic activities. However, based on our results and the scientific evidence presented above, we believe that the characterization and functional investigation of TPSB2 protein in the context of human lung adenocarcinoma might shed more light on the development or progression of this disease.

Palmitoylated proteins represent substrates of the cytoplasmic protein Acyl-protein thioesterase 1, the *LYPLA1* gene product (Zeidman et al., 2009; Hirano et al., 2009). In more detail, palmitoylation refers to the linkage of palmitic acid, a saturated fatty acid, to proteins through S-acylation (Zeidman et al., 2009). Although the exact role of LYPLA1 *in vivo* is not fully understood, it is important to mention that cycles of palmitoylation and depalmitoylation, the latter being catalyzed by LYPLA1, can have effects on signaling, protein stability, protein-protein interactions and protein-membrane association (Dekker et al., 2010; Zeidman et al., 2009). For example, palmitoylation plays a role in controlling localization and function of *Ras* proto-oncogene products (Dekker et al., 2010). In this regard, Dekker et al. hypothesized that inhibition of the thioesterase activity of LYPLA1 leads "to loss of normal localization by redistribution to all cellular membranes" of certain *Ras* gene products (Dekker et al., 2010). In order to test this hypothesis, Dekker et al. developed an LYPLA1 inhibitor and demonstrated that inhibition of this enzyme leads to a localization of *Ras* gene products to all membranes of Madin-Darby Canine Kidney (MDCK) epithelial cells (Dekker et al., 2010). Furthermore, the researchers observed a partial phenotypic reversion of MDCK-F3 cells, a Harvey murine sarcoma virus-transformed derivative of MDCK cells upon LYPLA1 inhibition (Dekker et al., 2010).

By making use of *www.genesapiens.org* we found that *LYPLA1* transcript levels are elevated in lung cancer tissues compared to non-neoplastic tissues of the respiratory system (see Supplemental Figure S3) (Kilpinen et al., 2008). These results indicate that differences on a transcript level do correspond with changes in enzymatic activities in case of Acyl-protein thioesterase 1 (see Figure 16). However, the elevated activity of LYPLA1 in the vast majority of lung adenocarcinoma biopsies investigated in this study and the regulatory influence of LYPLA1 on *Ras* proto-oncogene products suggest that Acyl-protein thioesterase 1 represents an interesting protein to be further investigated as a novel therapeutic target for human lung adenocarcinoma.

Lysophospholipase-like protein 1, the *LYPLAL1* gene product, exhibited elevated activities in the vast majority of investigated lung adenocarcinoma biopsies compared to matching non-neoplastic tissues. The function of LYPLAL1 remains to be determined, however, in accordance to our findings, *LYPLAL1* transcript levels were elevated in human lung cancer biopsies compared to normal tissues of the respiratory system as determined with *www.genesapiens.org* (see Supplemental Figure S3) (Kilpinen et al., 2008).

The serine hydrolase Isoamyl acetate-hydrolyzing esterase 1, the *IAH1* gene product, exhibited elevated activities in the vast majority of investigated malignant tissues. The function of IAH1 remains elusive, however, as determined with *www.genesapiens.org*, we found no differences in *IAH1* transcript levels in human lung cancer biopsies compared to normal tissues of the respiratory system, indicating that differences in transcript abundances do not necessarily correspond with changes in enzymatic activities (see Supplemental Figure S3) (Kilpinen et al., 2008).

Abhydrolase domain-containing protein 10 (ABHD10), an uncharacterized serine hydrolase with transcript levels not exhibiting significant differences in lung cancer versus normal tissues of the respiratory system as determined with *www.genesapiens.org*, also exhibited higher activities in the vast majority of lung adenocarcinoma biopsies investigated within this study (see Supplemental Figure S3) (Kilpinen et al., 2008). By making use of the Human Protein Atlas, a publicly accessible database that aims at comprehensively mapping protein expression and localization profiles in cancerous and normal tissues, we found that ABHD10 is expressed in a wide variety of human malignancies and healthy tissues (*www.proteinatlas.org*, February 18, 2011) (Uhlen et al., 2005; Berglund et al., 2008). More specifically, cytoplasmic ABHD10 expression has been reported in respiratory epithelial cells of the bronchus, pneumocytes and macrophages of the lung (*www.proteinatlas.org*, February 18, 2011) (Uhlen et al., 2005; Berglund et al., 2008). At this point it is important to mention that proteins expressed by a variety of cell types are suboptimal candidates for quantitative activity-based analysis, especially when heterogeneous specimens are analyzed. Unfortunately, due to the sensitivity of enzymatic activities to detergents, activity-based proteomics is not suited to be combined with methods like laser capture microdissection (LCM) for selective tumor tissue enrichment (see section Publications) (Collaud et al., 2010). To circumvent this limitation we recommend to assess tumor contents per surface and tumor viabilities on whole sections of frozen samples and exclude tissues with low tumor cell contents that additionally exhibit low tumor viability (see Supplemental Table S1).

In summary, TPSB2 protein, LYPLA1, LYPLAL1, IAH1 and ABHD10 exhibited decreased or elevated activities in lung adenocarcinoma biopsies compared to matching non-neoplastic tissues investigated within this study in an almost exclusive manner (see Figure 16). We therefore believe that these enzymes might play a general role in the pathogenesis of lung adenocarcinoma and that the functional characterization of LYPLAL1, IAH1 and ABHD10 will shed more light on the pathogenesis of lung cancer. We also want to point out that these results indicate that differences on a transcript level do not necessarily correspond with changes in enzymatic activities, thereby highlighting a major advantage of activity-based proteomics over conventional screening techniques.

4.2.3 Identification of ESD and ABHD11 activities as biomarker candidates for human lung adenocarcinoma

4.2.3.1 Esterase D (ESD)

The *ESD* gene maps close to the *RB1* gene locus 13q14, a genetic region that, when mutated, leads to the development of retinoblastoma (Wu et al., 2009; Parsam et al.,

2009). Due to the proximity of these two genes, *ESD* has successfully been employed as a genetic marker for retinoblastoma, a cancer of the eye that mostly affects children and that can be treated well if diagnosed early (Wu et al., 2009; Friend et al., 1986; Sparkes et al., 1983). Esterase D (ESD), a protein also known as S-formylglutathione hydrolase (FGH), is expressed in a wide variety of healthy and malignant tissues (*www.proteinatlas.org*, February 18, 2011) (Uhlen et al., 2005; Berglund et al., 2008). More specifically, ESD is expressed in the cytoplasm of respiratory epithelial cells of the bronchus, pneumocytes and macrophages of the lung (*www.proteinatlas.org*, February 18, 2011) (Uhlen et al., 2005; Berglund et al., 2008). Esterase D, the *ESD* gene product, is a glutathione thiol esterase that converts S-formylglutathione to glutathione and formate (van Straaten et al., 2009). ESD is involved in the detoxification pathway of formaldehyde in eukaryotes and prokaryotes, besides that little is known about its function (Wu et al., 2009; Eiberg and Mohr, 1986; van Straaten et al., 2009). Although no association between ESD activities and tumor aggressiveness has been reported in a comparable activity-based study investigating human breast cancer tissues, reduced ESD activities have been linked to the susceptibility of a variety of other pathological conditions including toxic liver cirrhosis, obesity and autism, indicating that reduced ESD activities play a role in the pathogenesis of a broad range of diseases (Jessani et al., 2005; Gonzalez et al., 2006).

In contradiction to our findings, we found no differences in *ESD* transcript levels in normal tissues of the respiratory system versus lung cancer biopsies by employing the publicly accessible human gene expression database *www.genesapiens.org* (Kilpinen et al., 2008). This observation supports previous findings that transcript levels and corresponding enzymatic activities do not necessarily correlate (see Figure 16 and Figure 21) (Jessani et al., 2005). This assumption is furthermore strengthened by the fact that *ESD* has been suggested as a reference gene for accurate normalization of gene expression profiling studies in NSCLC (Saviozzi et al., 2006). These findings indicate that the found activity differences might not be detectable on a transcript level.

As mentioned in section 3.3, the term "tumor grade" reflects the extend of tumor cell differentiation as assessed by a pathologist through microscopic inspection

(Suster, 2007). By making use of logistic regression, the results of this study revealed that the activity of Esterase D (ESD), also known as S-formylglutathione hydrolase (FGH), statistically significantly predicts the presence of high-grade lung adenocarcinomas ($p<0.05$, see section 3.4). Based on this statistically significant association we anticipate that upon further validation, ESD activities have the potential to develop into a useful tool for fine-tuning tumor grade assessment.

4.2.3.2 Abhydrolase domain-containing protein 11 (ABHD11)

The term "Abhydrolase" refers to a protein fold, namely the α/β hydrolase fold, that is common to many hydrolytic enzymes (Ollis et al., 1992). The core of these enzymes consists of eight β-sheets that are connected by α-helices (Ollis et al., 1992). These enzymes "have diverged from a common ancestor so as to preserve the arrangement of the catalytic residues" (Ollis et al., 1992).

The *ABHD11* gene is ubiquitously expressed and commonly deleted in Williams-Beuren syndrome, a developmental disorder with symptoms including mental retardation, overfriendliness and visuospatial impairment (Schubert, 2009; Merla et al., 2002). On a protein level, ABHD11 is expressed in a wide variety of healthy and malignant tissues and, more specifically, ABHD11 expression has been reported in pneumocytes and macrophages of the lung (*www.proteinatlas.org*, February 18, 2011) (Uhlen et al., 2005; Berglund et al., 2008). In a study conducted by Brentnall et al., researchers found that overexpression of ABHD11, Isoform 5, is associated with neoplastic progression in patients suffering from ulcerative colitis (UC), a disease with symptoms including open sores in the colon (Brentnall et al., 2009). Furthermore, Huang and colleagues reported that the expression of the *ABHD11* gene is correlated with sensitivity of 23 breast cancer cell lines to the multitargeted kinase inhibitor dasatinib (Huang et al., 2007). Although it is known that different splice isoforms of *ABHD11* exist, the biological function of ABHD11 still remains to be determined (Bachovchin et al., 2010; Schubert, 2009).

The results of this study revealed that the activity of the uncharacterized protein Abhydrolase domain-containing protein 11 (ABHD11), Isoform 1, predicts the development of distant metastases in a statistically significant model ($p<0.05$, see section 3.4). We furthermore observed that the mitochondrial protein ABHD11 exhibited a higher activity in the majority of malignant tissues, indicating that ABHD11 might play a general role in the pathogenesis of lung adenocarcinoma (Forner et al., 2006). Interestingly, by making use of *www.genesapiens.org*, we found higher *ABHD11* transcript levels in normal tissues of the respiratory system versus lung cancer biopsies, indicating that differences on a transcript level do not necessarily correspond to changes in enzymatic activities and that found activity differences might not be detectable on a transcript level (see Figure 16 and Figure 22) (Kilpinen et al., 2008).

We anticipate that the recent development of an ABHD11 specific inhibitor will enhance the functional characterization of ABHD11, which in turn might shed more light on the role of ABHD11 in cancer in general and in lung adenocarcinoma in particular (Bachovchin et al., 2010). Furthermore, based on the statistically significant association between the development of distant metastases and ABHD11 activities, we anticipate that ABHD11 activities, alone or in combination with additional molecular characteristics, have the potential to develop into molecular predictors with a reliable clinical significance.

4.2.4 The potential of enzymatic activities as clinical biomarkers: chances and limitations

The results of this study provide evidence that SH activities bear significant prediction abilities for human lung adenocarcinoma and therefore represent attractive molecular characteristics that can potentially be employed as clinical biomarkers.

In our view, the successful implementation of enzymatic activities as clinical biomarkers is only impaired by technical aspects, i.e. although the implemented workflow represents a relatively fast and robust approach for biomarker discovery and partially biomarker validation studies, the presented workflow is not well suited to be employed as a diagnostic tool in a clinical laboratory working on a daily basis. Alternatively, we anticipate that previously described, easy-to-use microplate arrays may represent high-throughput tools that will circumvent this limitation (Sieber and

Cravatt, 2006; Sieber et al., 2004). However, one aspect that highlights the potential of enzymatic activities in general and of SH activities in particular to be employed as clinical biomarkers is discussed in the following: all SHs investigated in this study represent "druggable" biomolecules and can therefore be seen as potential targets for novel therapeutic interventions (Russ and Lampel, 2005). Thereby it should be noted that almost 50% of all marketed small-molecule drugs aim at targeting enzymatic activities (Hopkins and Groom, 2002). However, one frequent argument is that selective inhibitors for the majority of SHs still need to be developed (Bachovchin et al., 2010). In order identify selective SH inhibitors, several methodologies relying on activity-based proteomics have been presented so far (Bachovchin et al., 2009; Bachovchin et al., 2010). Thereby, differences in binding kinetics of unselective SH inhibitors, FP-biotin for example, and binding kinetics of inhibitor candidates derived from a compound library, are exploited. Several inhibitors selectively targeting SHs have already been identified by employing this strategy with methodologies that became known as competitive activity-based protein profiling (competitive ABPP) or fluopol-ABPP (Bachovchin et al., 2009; Bachovchin et al., 2010).

However, in this study we have presented an advanced activity-based proteomics platform for clinical biomarker discovery. We have provided evidence for the potential of a so far widely unexploited class of molecular characteristics, namely enzymatic activities, to serve as biomarkers for human lung adenocarcinoma. Although data validation with an independent and increased sample set will be required to perform clinically more relevant statistical analyses, we anticipate that the activities of Esterase D (ESD) as well as of Abhydrolase domain-containing protein 11 (ABHD11) have the potential to develop into molecular predictors with a reliable clinical significance. Finally, due to the versatility and unique properties of the technique, we conclude that activity-based proteomics complements existing biomarker screening techniques and holds great promise in the expanding field of clinical biomarker discovery.

5. Conclusion

Prognosis for patients suffering from lung cancer is currently mostly determined based on the extension of disease at diagnosis. Thereby, it has become evident that predicted and real outcomes can vary significantly, even for patients with the same stage of disease. In order to refine prognosis, novel biomarkers with a reliable predictive significance are clearly needed.

In this PhD project the potential of SH activities as clinical biomarkers was evaluated. The rationale for focusing on SH activities was that i) transcript and corresponding protein levels do not necessarily correlate with activity states and conventional biomarker discovery strategies might therefore fail to detect crucial changes in enzymatic activities caused by posttranslational events during tumor progression and treatment response, ii) the serine hydrolase superfamily comprises a large and diverse repertoire of enzymes that make up 1% of the human proteome, iii) serine hydrolases have extensively been studied in an activity-based manner due to the unique molecular characteristics of their active site (see section 1.6) and iv) members of the serine hydrolase superfamily have previously been linked to lung cancer (Gygi et al., 1999b; Alaiya et al., 2000; Sieber and Cravatt, 2006; Sieber et al., 2006; Jessani et al., 2005; Simon and Cravatt, 2010; Meyer et al., 2004; Zelvyte et al., 2004).

Activity-based proteomics has already been introduced as a high-throughput platform for the investigation of SH activities in human specimens (Jessani et al., 2005). However, due to i) the uncoupled qualitative and quantitative analysis with unequal sensitivity levels and ii) their interdependence we conclude that this approach is not well suited to be employed in clinical biomarker discovery studies investigating human lung adenocarcinoma biopsies. Therefore, an advanced, solely LC-MS/MS based and label-free platform for the investigation of SH activities in complex proteomes has been implemented. The methodology is especially well suited to be employed in clinical biomarker discover studies due to the following reasons: i) simultaneous identification and quantification of enzymatic activities, ii) in case malignant and matching non-neoplastic tissues are available, SH activity signatures can be normalized on an individual basis, iii) good repeatability and reproducibility, iv) label-free quantification is a cost-effective alternative to quantification strategies employing isotopically labeled activity-based probes, v) serine hydrolase activities derived from virtually all kinds of tissues and body fluids can be investigated and vi) the methodology is in principle applicable to every enzyme class for which activity-based probes have been developed. However, limitations of this workflow include i) long analysis times, especially when high numbers of replicates are investigated and ii) the high proteome consumption as a consequence thereof. Based on these considerations we conclude that the implemented workflow is well suited to be employed in biomarker discovery and partially in biomarker validation studies, however, we anticipate that other technological solutions like microplate arrays will be required to implement an activity-based diagnostic test in a clinical laboratory working on a daily basis (see Figure 23) (Sieber and Cravatt, 2006; Sieber et al., 2004).

Within this study, 40 pairs of malignant and matching non-neoplastic lung tissues were investigated for SH activities. The results of this study revealed that TPSB2 protein, LYPLA1, LYPLAL1, IAH1, ABHD10 and ABHD11 exhibited decreased or elevated activities in lung adenocarcinoma biopsies compared to matching non-neoplastic tissues in an almost exclusive manner. These results indicate that the enzymes mentioned above play a general role in the pathogenesis of lung adenocarcinoma and that further investigation and characterization of these proteins will shed more light on the development and / or progression of this disease.

Interestingly, on a transcript level, *TPSB2*, *IAH1*, and *ABHD10* did not show any differences in lung cancer biopsies compared to normal tissues of the respiratory system as determined with the publicly accessible gene expression database *www.genesapiens.org* (Kilpinen et al., 2008). Differences on a transcript level compared to differences on an activity level were even contradictory in case of ABHD11 (Kilpinen et al., 2008). These observations indicate that the found SH activity differences might not be detectable on a transcript level.

Furthermore, the activities of ESD predicted the presence of high-grade tumors and the activities of ABHD11 predicted the development of distant metastases, both in a statistically significant model. Although data validation with an increased sample set will be required to perform clinically more relevant statistical analysis, we anticipate that ESD and / or ABHD11 activities have the potential to develop into molecular predictors with a reliable clinical significance.

Based on this PhD project, a feasible platform for the identification of SH activities as potential biomarker candidates is now available. We suggest that the implemented methodology is ideally employed in the discovery phase of the activity-based biomarker development process as illustrated in Figure 23. Furthermore, the results obtained in this PhD project provide evidence that SH activities, particularly the activities of ESD and ABHD11, two enzymes that have previously not been associated with NSCLC, bear predictive potential for human lung adenocarinoma and therefore represent attractive molecular characteristics to be employed as clinical biomarkers. Finally, based on the results of this work we believe that activity-based proteomics complements existing methodologies employed in the seek for disease biomarkers and holds great promise in the expanding field of biomarker discovery.

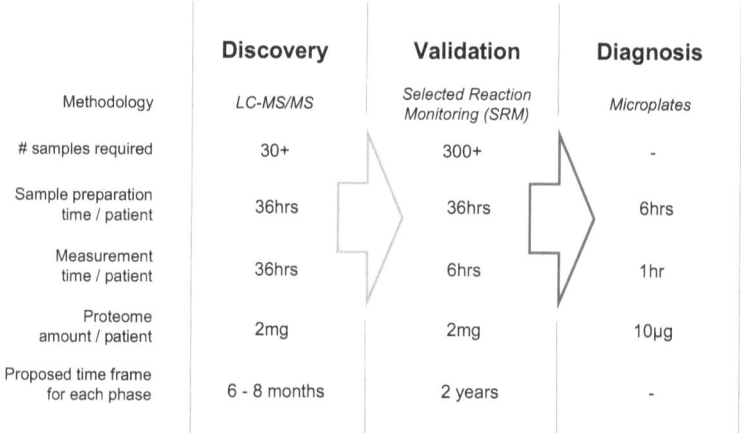

Figure 23 - **Activity-based biomarker development.** After the discovery phase, biomarker candidates are validated using SRM. Although the number of replicates analyzed can be minimized in the validation phase due to excellent reproducibility inherent to SRM-based experiments, sample preparation time per patient does not change and therefore remains a time-intense procedure (Lange et al., 2008). For biomarker validation, we highly recommend to perform statistical power analysis to determine the optimal number of samples required. After biomarkers have successfully been validated, we anticipate that due to i) the low sample preparation and measurement time, ii) the low proteome consumption and iii) the capability of analyzing enzymatic activities of many patients in parallel, microplates represent valid tools to be employed for activity-based diagnostics (Sieber and Cravatt, 2006). Time frames were calculated for one full time employee.

Supplementary Data

Supplemental Figure S1 - Proteome quality assessment and activity-based probe labeling efficiencies. Prior to activity-based probe incubation, proteome qualities were assessed through protein separation by 1D-SDS-PAGE and subsequent Coomassie Blue staining. Activity-based probe labeling efficiencies were monitored through parallel incubation of the SH trypsin (Promega) and analysis by Western Blot (TU = malignant tissue, NO = corresponding non-neoplastic tissue).

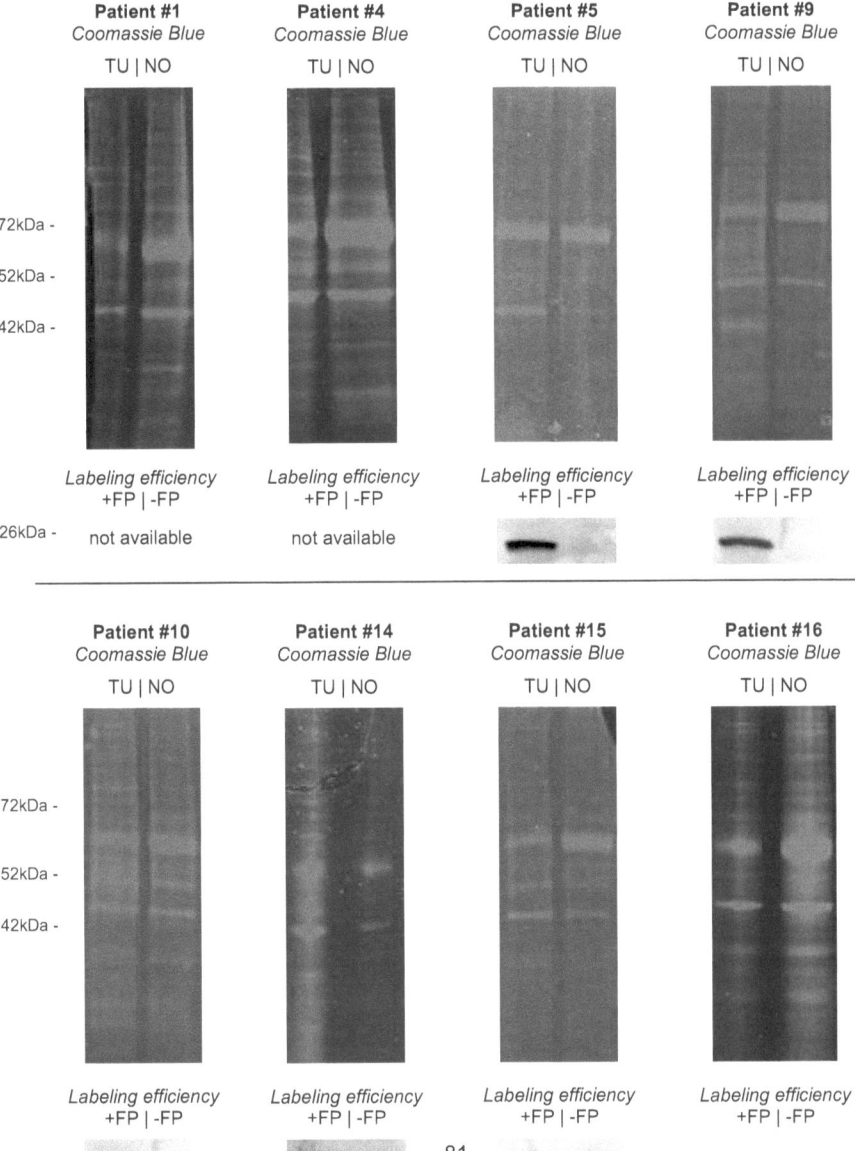

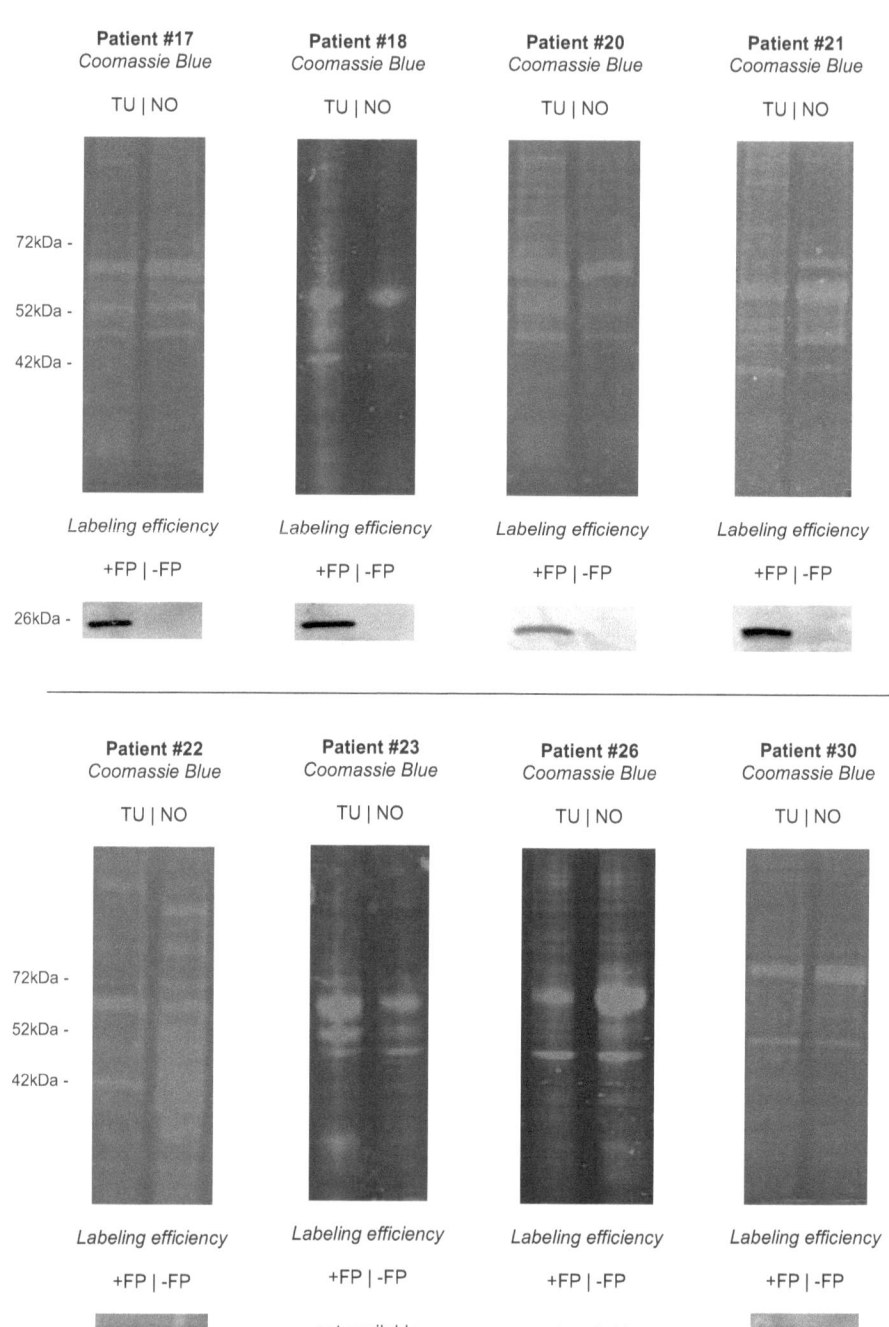

Patient #31	Patient #32	Patient #33	Patient #34
Coomassie Blue	*Coomassie Blue*	*Coomassie Blue*	*Coomassie Blue*
TU | NO	TU | NO	TU | NO	TU | NO

72kDa -
52kDa -
42kDa -

Labeling efficiency	*Labeling efficiency*	*Labeling efficiency*	*Labeling efficiency*
+FP | -FP	+FP | -FP	+FP | -FP	+FP | -FP

26kDa -

Patient #35	Patient #36	Patient #38	Patient #39
Coomassie Blue	*Coomassie Blue*	*Coomassie Blue*	*Coomassie Blue*
TU | NO	TU | NO	TU | NO	TU | NO

72kDa -
52kDa -
42kDa -

Labeling efficiency	*Labeling efficiency*	*Labeling efficiency*	*Labeling efficiency*
+FP | -FP	+FP | -FP	+FP | -FP	+FP | -FP
	not available	not available	not available

26kDa -

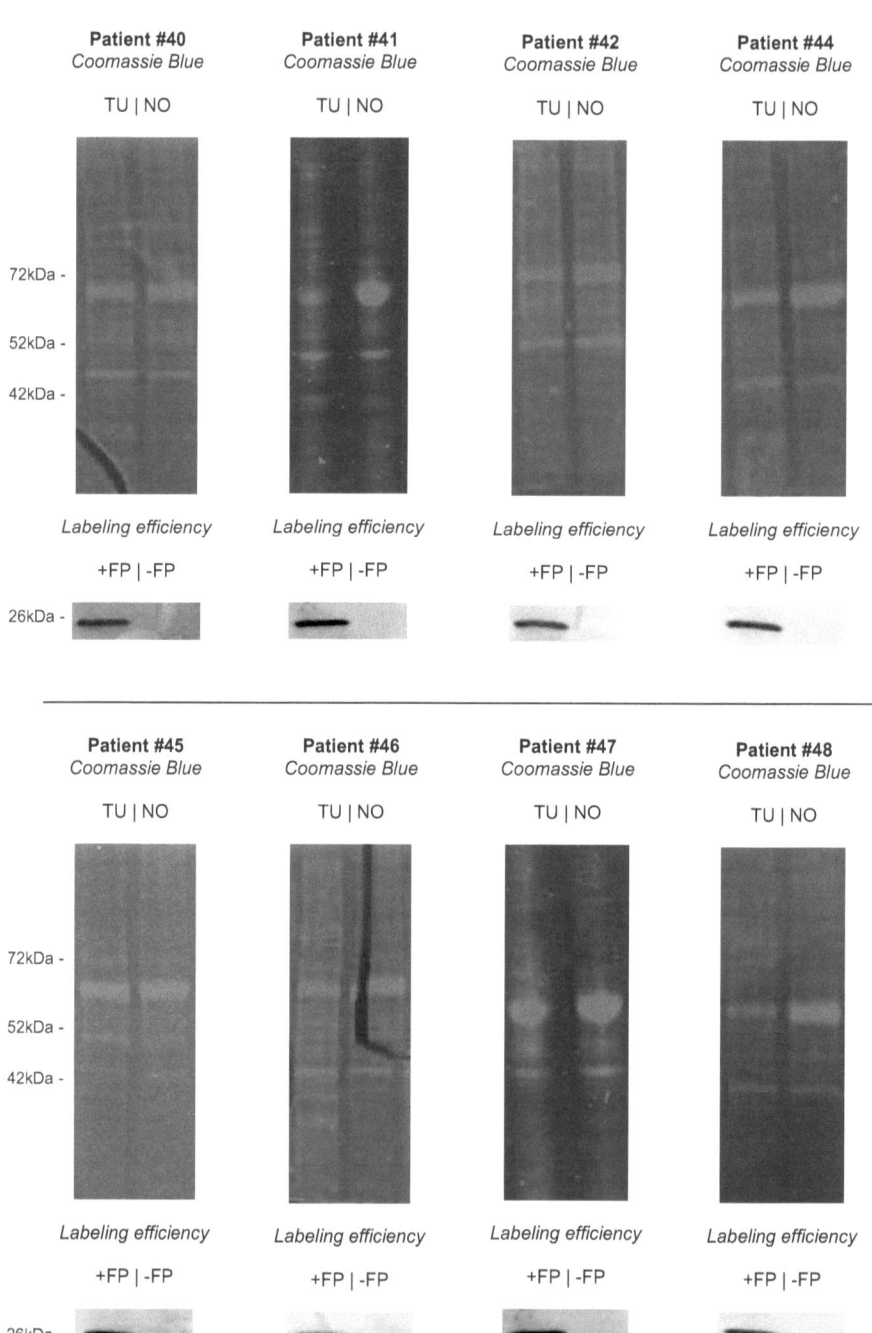

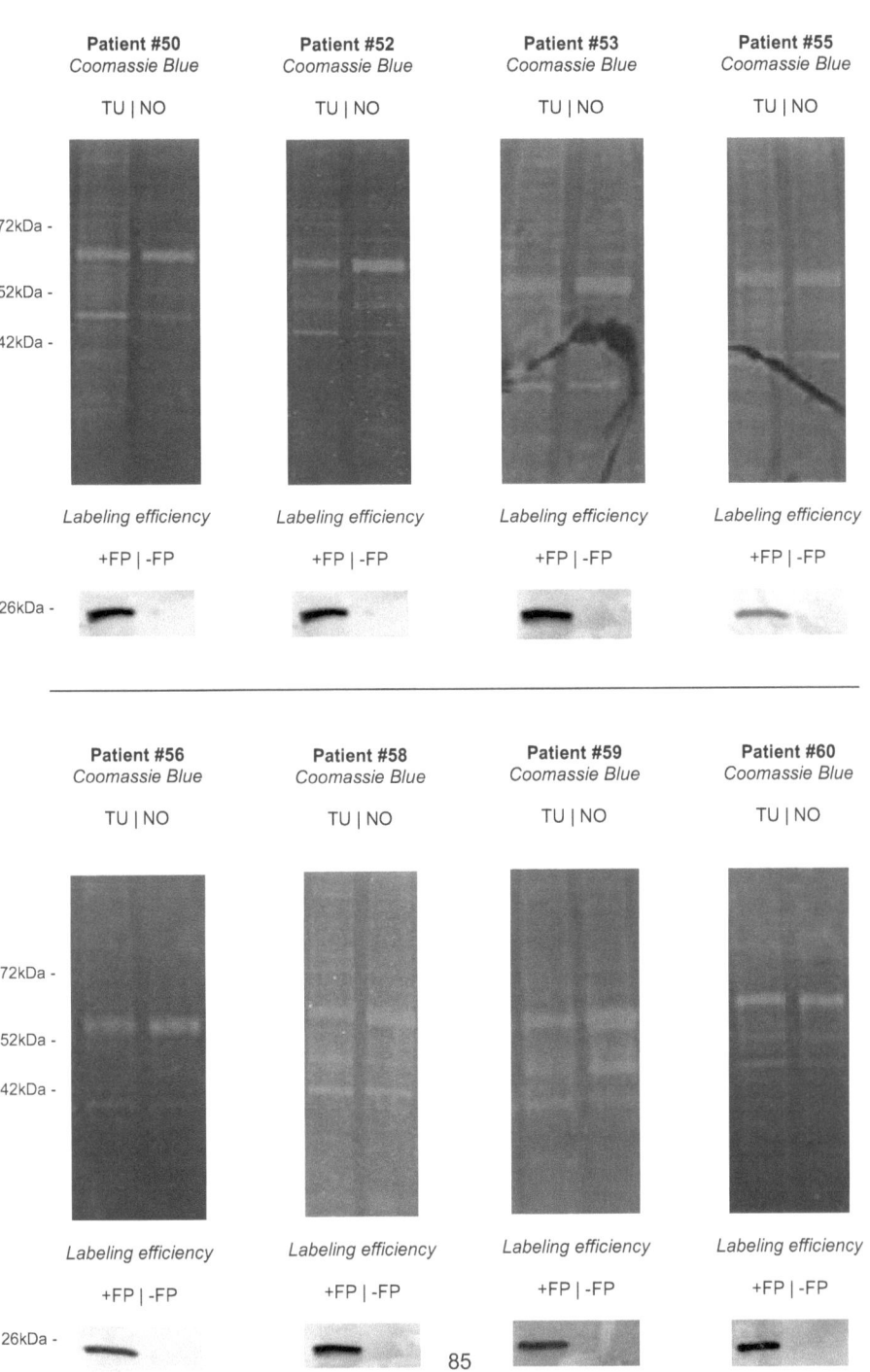

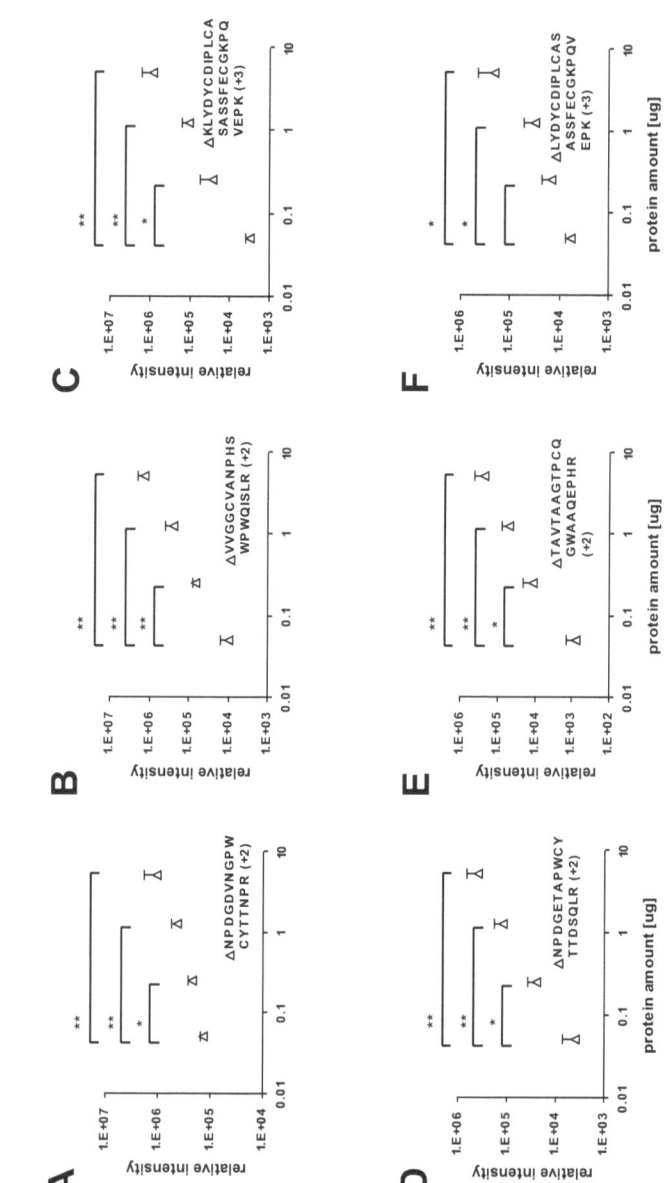

Supplemental Figure S2 A - L) Relative quantification of low intense peptide ions matching to the FP-1 labeled SH murine Plasmin in a complex tumor proteome. Peptides with relative intensities as low as 1E+02 exhibit a good correlation between relative intensity and protein amount. Samples were measured in biochemical triplicates (*: $p<0.05$; **: $p<0.005$; unpaired, two sided Student's t-test with log2 transformed values).

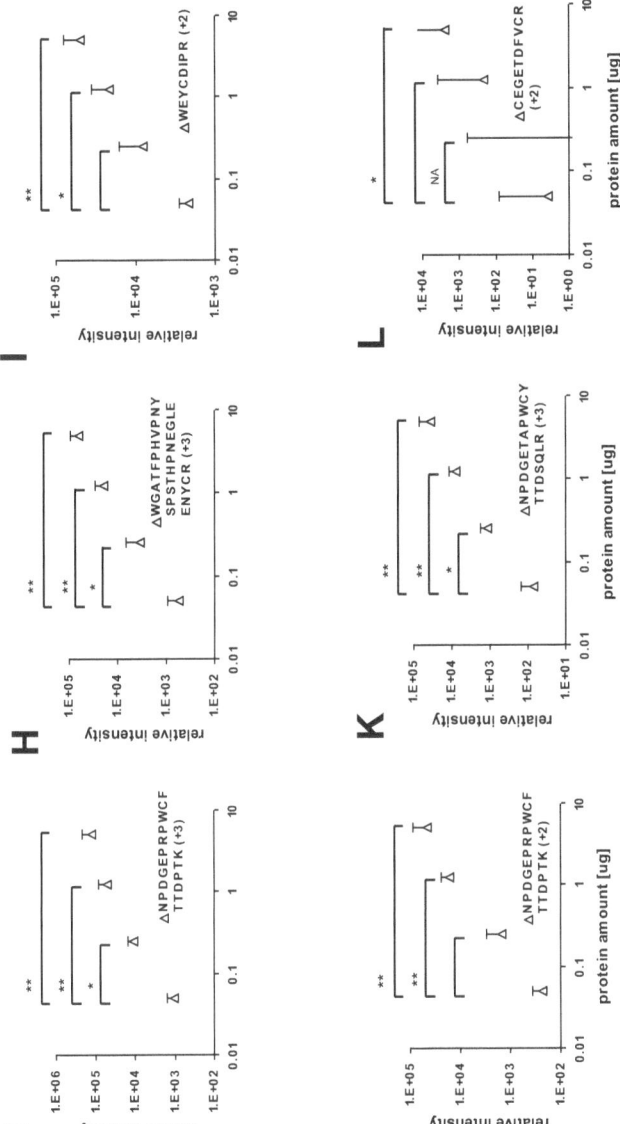

Supplemental Figure S3 - *LYPLA1*, *LYPLAL1*, *IAH1*, *ABHD10* and *TPSB2* transcript levels in healthy and malignant tissues of the respiratory system. S3 A - B) A slightly higher level of *LYPLA1* and *LYPLAL1* expression can be observed in malignant tissues. S3 C - E) On a transcript level, no differences of *IAH1*, *ABHD10* and *TPSB2* expression can be observed in malignant compared to healthy tissues of the respiratory system. The plots were downloaded from www.*genesapiens.org* on February 18, 2011.

A

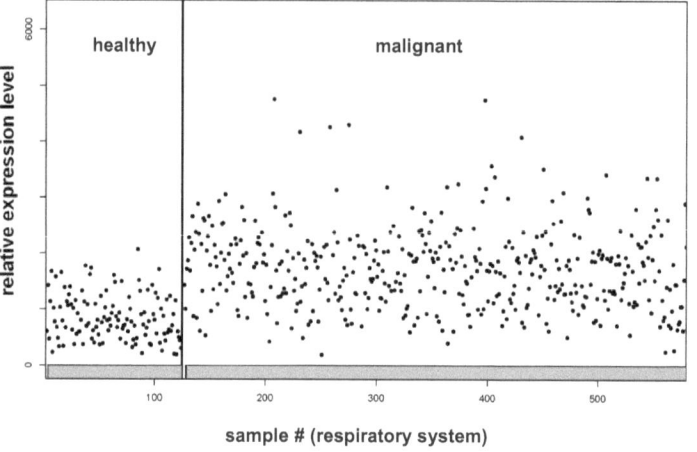

B

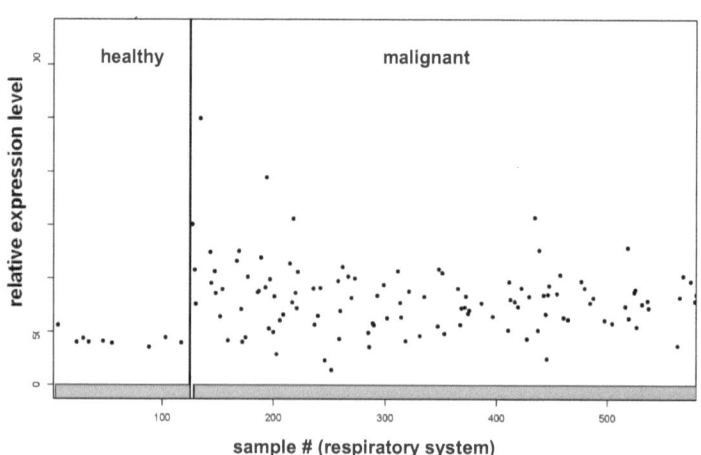

C

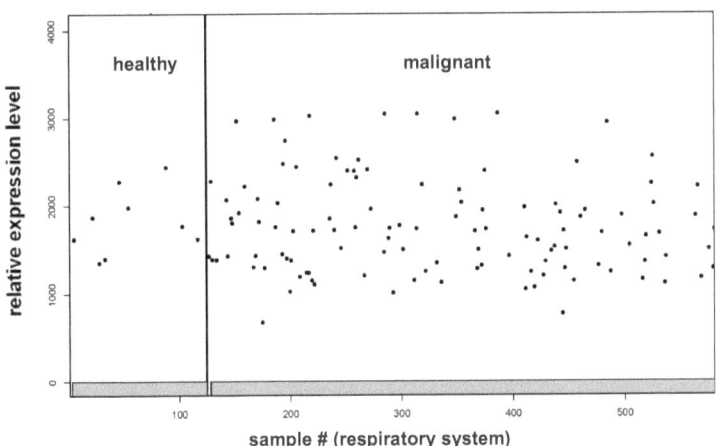

Isoamyl acetate-hydrolyzing esterase 1 (*IAH1*), transcript

D

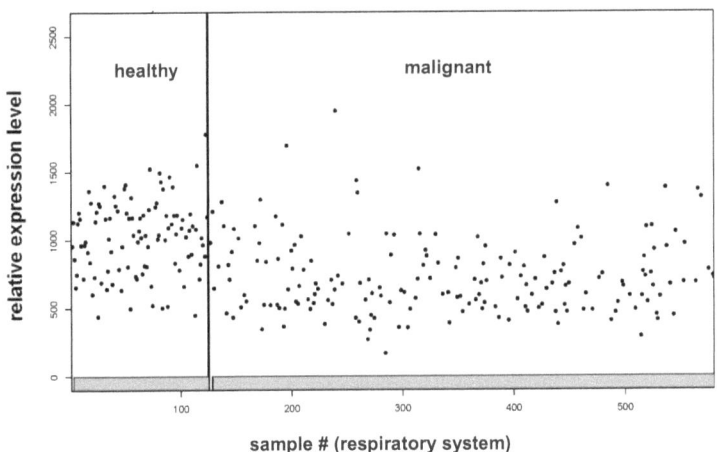

Abhydrolase domain-containing protein 10 (*ABHD10*), transcript level

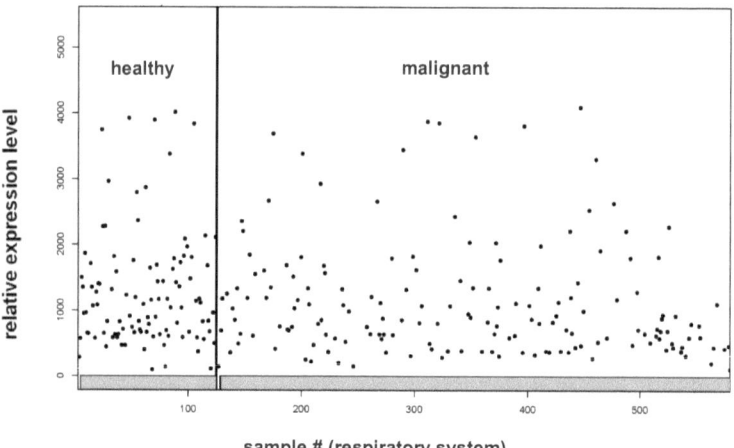

Supplemental Table S1 - Comprehensive clinical follow-up data of participating individuals.

	Patients characteristics							Tumor cell content			
ID #	Surgery	TNM stage[1]	Resection	Lymph node status	Grade	Distant metastases	Gender	% tumor	% stroma	% necrosis	% normal
1	lobectomy	IA	R0	N0	G2	-	F	60	10	5	5
4	lobectomy	IA	R0	N0	G3	-	F	60	35	0	5
5	lobectomy	IA	R0	N0	G2	-	M	30	70	0	0
9	lobectomy	IB	R0	N0	G3	-	M	80	20	0	0
10	wedge resection	IB	R0	N0	G2	-	M	60	30	10	0
14	lobectomy	IB	R0	N0	G3	-	M	70	25	5	0
15	lobectomy	IB	R0	N0	G3	-	M	50	45	-	5
16	lobectomy	IB	R0	N0	G1	-	M	60	40	-	-
17	lobectomy	IB	R0	N0	G3	-	M	60	40	-	-
18	lobectomy	IB	R0	N0	G3	YES (stomach)	M	70	30	-	-
20	lobectomy	IB	R0	N0	G2	-	W	50	35	5	10
21	lobectomy	IIA	R0	N1	G2	NA[2]	W	30	70	-	-
22	lobectomy	IIB	R0	N1	G1	-	W	40	40	20	-
23	pneumonectomy	IIIA	R0	N2	G1	YES (liver)	M	90	10	-	-
26	lobectomy	IIIA	R0	N2	G3	-	W	60	40	-	-
30	lobectomy	IA	R0	N0	G2	-	M	70	30	-	-
31	lobectomy	IA	R0	N0	G2	-	M	80	20	-	10

[1] Classification according to 6th TNM-Classification of Malignant Tumors (2002).
[2] NA: not available

ID #	Surgery	TNM stage	Resection	Lymph node status	Grade	Distant metastases	Gender	% tumor	% stroma	% necrosis	% normal
32	lobectomy	IA	R0	N0	G2	-	M	60	40	-	-
33	lobectomy	IB	R0	N1	G2	YES (brain, bone)	M	40	60	-	-
34	lobectomy	IB	R0	N0	G2	YES (brain)	M	70	30	-	-
35	lobectomy	IB	R1	N0	-	excluded (R1)	M	50	45	5	-
36	lobectomy	IIA	R0	N1	G3	YES (brain)	W	80	20	-	-
38	lobectomy	IIB	R1	N1	G2	excluded (R1)	M	70	30	-	-
39	lobectomy	IIB	R0	N1	G1	YES (brain, lung)	M	90	10	-	-
40	pneumonectomy	IIIA	R0	N2	G2	YES (lung)	M	60	35	5	-
41	lobectomy	IIIA	R0	N2	G3	-	M	90	10	-	-
42	lobectomy	IIIA	R0	N2	G3	-	M	80	20	-	-
44	lobectomy	IIIA	R0	N2	G3	YES (bone)	W	50	50	-	-
45	lobectomy	IIIA	R1	N1	G3	excluded (R1)	W	30	30	40	-
46	pneumonectomy	IIIA	R0	N2	G3	-	W	80	20	-	-
47	pneumonectomy	IIIA	R0	N1	G3	YES (brain, lung)	W	80	20	-	-
48	pneumonectomy	IIIA	R1	N2	G3	excluded (R1)	W	50	30	-	20
50	wedge resection	IIIA	R0	N2	G3	YES (liver, bone)	M	50	50	-	-
52	pneumonectomy	IIIB	R1	N2	G3	excluded (R1)	M	50	50	-	-
53	lobectomy	IIIB	R0	N2	G2	-	W	20	80	-	-
55	lobectomy	IV	R1	N0	G3	excluded (R1)	M	70	30	-	-
56	pneumonectomy	IV	R0	N3	G3	excluded (M1 at diagnosis)	M	40	60	-	-
58	lobectomy	IV	R0	N2	G3	excluded (M1 at diagnosis)	M	50	30	-	20
59	lobectomy	IV	R1	N0	G2	excluded (R1)	W	50	40	-	10
60	wedge resection	IV	R1	N3	G3	excluded (R1)	W	70	30	-	-

Supplemental Table S2 - Complete list of SHs identified in human lung adenocarcinoma biopsies and the human lung adenocarcinoma cell line CaLu-3.

Identified Serine Hydrolases

#	Full protein name	UniProtKB / Swiss-Prot ID	Gene symbol
1	1-O-acylceramide synthase	Q8NCC3	LYPA3
2	Abhydrolase domain-containing protein 10, mitochondrial precursor	Q9NUJ1	ABHDA
3	Abhydrolase domain-containing protein 11	Q8NFV4	ABHDB
4	Abhydrolase domain-containing protein 6	Q9BV23	ABHD6
5	Acetylcholinesterase	P22303	ACHE
6	Acylamino-acid-releasing enzyme	P13798	ACPH
7	Acyl-coenzyme A thioesterase 1	Q86TX2	ACOT1
8	Acyl-coenzyme A thioesterase 2, mitochondrial	P49753	ACOT2
9	Acyloxyacyl hydrolase	P28039	AOAH
10	Acyl-protein thioesterase 1	O75608-1	LYPA1
11	Acyl-protein thioesterase 2	O95372	LYPA2
12	Arylacetamide deacetylase	P22760	AADAC
13	Arylacetamide deacetylase-like 1 (KIAA1363)	Q6PIU2	ADCL1
14	BAT5 (HLA-B associated transcript 5)	A2BEY3	BAT5
15	Bile salt-activated lipase	P19835-1	CEL
16	Carboxylesterase 2	O00748	CES2
17	Carboxylesterase 3	Q6UWW8	CES3
18	Cathepsin G	P08311	CTSG
19	cDNA FLJ34625 fis	B3KRN4	-
20	Cholinesterase	P06276	BCHE
21	Dipeptidyl peptidase 4	P27487	DPP4
22	Dipeptidyl peptidase 8	Q6V1X1-1	DPP8
23	Dipeptidyl peptidase 9	Q86TI2	DPP9
24	Dipeptidyl-peptidase 2	Q9UHL4	DPP2

#	Full protein name	UniProtKB / Swiss-Prot ID	Gene symbol
25	Fatty acid synthase	P49327	FAS
26	Fatty-acid amide hydrolase 1	O00519	FAAH
27	Granzyme A	P12544	GZMA
28	Granzyme K	P49863	GZMK
29	IAH1 protein	Q05D21	IAH1
30	Isoamyl acetate-hydrolyzing esterase 1 homolog	Q2TAA2	IAH1
31	Kallikrein-6	Q92876-1	KLK6
32	Leukocyte elastase / Neutrophile elastase	P08246	ELANE
33	Lipoprotein lipase	P06858	LPL
34	Liver carboxylesterase 1	P23141-1	EST1
35	Lysophospholipase-like protein 1	Q5VWZ2-1	LYPL1
36	Lysosomal protective protein precursor (Cathepsin A)	P10619	PPGB
37	Lysosomal Pro-X carboxypeptidase precursor	P42785	PCP
38	Monoacylglycerol lipase	B2ZGL7	MGLL
39	Monoglyceride lipase	Q6IBG9	MGLL
40	Myeloblastin	P24158	PRTN3
41	Neuropathy target esterase	Q8IY17-1	PNPLA6
42	Ovarian cancer-associated gene 2 protein	Q8WZ82	OVCA2
43	Palmitoyl-protein thioesterase 2	A2ABN6	PPT2
44	Patatin-like phospholipase domain-containing protein 4	P41247	PLPL4
45	Patatin-like phospholipase domain-containing protein 7	PLPL7	PNPLA7
46	Plasma kallikrein	P03952	KLKB1
47	Plasminogen	P00747	PLG
48	Platelet-activating factor acetylhydrolase	Q13093	PLA2G7
49	Platelet-activating factor acetylhydrolase 2, cytoplasmic	Q99487	PAFA2
50	Platelet-activating factor acetylhydrolase IB subunit beta	P68402	PA1B2
51	Platelet-activating factor acetylhydrolase IB subunit gamma	Q15102	PA1B3
52	Presenilins-associated rhomboid-like protein, mitochondrial	Q9H300	PARL

#	Full protein name	UniProtKB / Swiss-Prot ID	Gene symbol
53	Probable arylformamidase	A2RUB3	AFMID
54	Probable arylformamidase (AFMID protein)	A2RUB3	AFMID
55	Probable serine carboxypeptidase CPVL precursor	Q9H3G5	CPVL
56	Prolyl endopeptidase	P48147	PPCE
57	Prolyl endopeptidase like	Q4J6C6-1	PREPL
58	Prolylcarboxypeptidase (Angiotensinase C)	A8MU24	PRCP
59	Prostasin	Q16651	PRSS8
60	Protease serine 4 isoform B / Trypsin-3	P35030	PRSS3
61	Proteasome subunit alpha type 6	P60900	PSA6
62	Proteasome subunit alpha type-5	P28066	PSA5
63	Proteasome subunit beta type-2	P49721	PSB2
64	Proteasome subunit beta type-4	P28070	PSMB4
65	Proteasome subunit beta type-5	P28074	PSB5
66	Proteasome subunit beta type-8	P28062	PSMB8
67	Proteasome subunit beta type-9	P28065-1	PSMB9
68	Protein phosphatase methylesterase 1	Q9Y570	PPME1
69	PRSS1 protein	Q3SY19	PRSS1
70	Putative hydrolase RBB9	O75884-1	RBB9
71	Putative uncharacterized protein ENSP00000381469	A8MXM1	A8MXM1
72	Retinoid-inducible serine carboxypeptidase	Q9HB40-1	SCPEP1
73	Seprase	Q12884-1	SEPR
74	Serine beta-lactamase-like protein LACTB, mitochondrial	P83111-1	LACTB
75	Serine protease 1-like protein 1	Q6UWY2	PRSSL1
76	S-formylglutathione hydrolase	P10768	ESD
77	Sialate O-acetylesterase	Q9HAT2-1	SIAE
78	Sn1-specific diacylglycerol lipase beta	Q8NCG7	DGLP
79	Tissue-type plasminogen activator	P00750	PLAT
80	TPSB2 protein	Q6NZY1	TPSB2

#	Full protein name	UniProtKB / Swiss-Prot ID	Gene symbol
81	TypeII 3a hydroxysteroid dehydrogenase variant	Q2XPP3	Q2XPP3
82	Urokinase-type plasminogen activator	P00749	PLAU

Publications

This study has been conducted at and supported by the University Hospital Zurich, Switzerland, the Institute of Molecular Systems Biology, ETH Zurich, Switzerland and the Functional Genomics Center Zurich, Switzerland.

Data presented in this publication have been published in the following journals:

Activity-based proteomics: identification of ABHD11 and ESD activities as potential biomarkers for human lung adenocarcinoma. Wiedl T, Arni S, Roschitzki B, Grossmann J, Collaud S, Soltermann A, Hillinger S, Aebersold R, Weder W. J Proteomics. 2011 Sep 6;74(10):1884-94. Epub 2011 May 9.

KRAS Mutation Is Associated with Elevated Myeloblastin Activity in Human Lung Adenocarcinoma. Wiedl T, Collaud S, Hillinger S, Arni S, Burgess C, Kroll W, Schraml P, Soltermann A, Moch H, Weder W. Cancer Genomics Proteomics. 2012 Jan;9(1):51-4.

References

Aebersold R & Mann M. (2003) Mass spectrometry-based proteomics. *Nature* 422: 198-207.

Alaiya AA, Franzen B, Auer G & Linder S. (2000) Cancer proteomics: from identification of novel markers to creation of artifical learning models for tumor classification. *Electrophoresis* 21: 1210-1217.

Anderson NL & Anderson NG. (1998) Proteome and proteomics: new technologies, new concepts, and new words. *Electrophoresis* 19: 1853-1861.

Arai H, Koizumi H, Aoki J & Inoue K. (2002) Platelet-activating factor acetylhydrolase (PAF-AH). *J Biochem* 131: 635-640.

Atkinson AJ, Colburn WA, G. DV, DeMets DL, Downing GJ, Hoth DF, Oates JA, Peck CC, Schooley RT, Spilker BA, Woodcock J & Zeger SL. (2001) Biomarkers and surrogate endpoints: Preferred definitions and conceptual framework. *Clinical Pharmacology & Therapeutics*: 89-95.

Bachovchin DA, Brown SJ, Rosen H & Cravatt BF. (2009) Identification of selective inhibitors of uncharacterized enzymes by high-throughput screening with fluorescent activity-based probes. *Nat Biotechnol* 27: 387-394.

Bachovchin DA, Ji T, Li W, Simon GM, Blankman JL, Adibekian A, Hoover H, Niessen S & Cravatt BF. (2010) Superfamily-wide portrait of serine hydrolase inhibition achieved by library-versus-library screening. *Proc Natl Acad Sci U S A* 107: 20941-20946.

Balk SP, Ko YJ & Bubley GJ. (2003) Biology of prostate-specific antigen. *J Clin Oncol* 21: 383-391.

Barletta JA, Yeap BY & Chirieac LR. (2010) Prognostic significance of grading in lung adenocarcinoma. *Cancer* 116: 659-669.

Bean J, Brennan C, Shih JY, Riely G, Viale A, Wang L, Chitale D, Motoi N, Szoke J, Broderick S, Balak M, Chang WC, Yu CJ, Gazdar A, Pass H, Rusch V, Gerald W, Huang SF, Yang PC, Miller V, Ladanyi M, Yang CH & Pao W. (2007) MET amplification occurs with or without T790M mutations in EGFR mutant lung tumors with acquired resistance to gefitinib or erlotinib. *Proc Natl Acad Sci U S A* 104: 20932-20937.

Berglund L, Bjorling E, Oksvold P, Fagerberg L, Asplund A, Szigyarto CA, Persson A, Ottosson J, Wernerus H, Nilsson P, Lundberg E, Sivertsson A, Navani S, Wester K, Kampf C, Hober S, Ponten F & Uhlen M. (2008) A genecentric Human Protein Atlas for expression profiles based on antibodies. *Mol Cell Proteomics* 7: 2019-2027.

Blackstock WP & Weir MP. (1999) Proteomics: quantitative and physical mapping of cellular proteins. *Trends Biotechnol* 17: 121-127.

Blum M. (2005) Clinical Presentation of Lung Cancer. In: Shields TW (ed) *General Thoracic Surgery*. Philadelphia, 1508-1517.

Bories D, Raynal MC, Solomon DH, Darzynkiewicz Z & Cayre YE. (1989) Down-regulation of a serine protease, myeloblastin, causes growth arrest and differentiation of promyelocytic leukemia cells. *Cell* 59: 959-968.

Brambilla C, Fievet F, Jeanmart M, de Fraipont F, Lantuejoul S, Frappat V, Ferretti G, Brichon PY & Moro-Sibilot D. (2003) Early detection of lung cancer: role of biomarkers. *Eur Respir J Suppl* 39: 36s-44s.

Brambilla EaL, Sylvie. (2008) Histopathology of lung tumors. In: Hanson H (ed) *Textbook of lung cancer*. second ed. London, UK: Informa Healthcare, 61-74.

Brentnall TA, Pan S, Bronner MP, Crispin DA, Mirzaei H, Cooke K, Tamura Y, Nikolskaya T, Jebailey L, Goodlett DR, McIntosh M, Aebersold R, Rabinovitch PS & Chen R. (2009) Proteins That Underlie Neoplastic Progression of Ulcerative Colitis. *Proteomics Clin Appl* 3: 1326-1337.

Broers JL, Viallet J, Jensen SM, Pass H, Travis WD, Minna JD & Linnoila RI. (1993) Expression of c-myc in progenitor cells of the bronchopulmonary epithelium and in a large number of non-small cell lung cancers. *Am J Respir Cell Mol Biol* 9: 33-43.

Cai YD, Zhou GP, Jen CH, Lin SL & Chou KC. (2004) Identify catalytic triads of serine hydrolases by support vector machines. *J Theor Biol* 228: 551-557.

Carlini MJ, Dalurzo MC, Lastiri JM, Smith DE, Vasallo BC, Puricelli LI & Lauria de Cidre LS. (2010) Mast cell phenotypes and microvessels in non-small cell lung cancer and its prognostic significance. *Hum Pathol* 41: 697-705.

Chakravarty B, Gu Z, Chirala SS, Wakil SJ & Quiocho FA. (2004) Human fatty acid synthase: structure and substrate selectivity of the thioesterase domain. *Proc Natl Acad Sci U S A* 101: 15567-15572.

Chang PA, Chen YY, Qin WZ, Long DX & Wu YJ. (2011) The role of cell cycle-dependent neuropathy target esterase in cell proliferation. *Mol Biol Rep* 38: 123-130.

Chen CH. (2004) Platelet-activating factor acetylhydrolase: is it good or bad for you? *Curr Opin Lipidol* 15: 337-341.

Chen G, Gharib TG, Wang H, Huang CC, Kuick R, Thomas DG, Shedden KA, Misek DE, Taylor JM, Giordano TJ, Kardia SL, Iannettoni MD, Yee J, Hogg PJ, Orringer MB, Hanash SM & Beer DG. (2003) Protein profiles associated with survival in lung adenocarcinoma. *Proc Natl Acad Sci U S A* 100: 13537-13542.

Chen WT & Kelly T. (2003) Seprase complexes in cellular invasiveness. *Cancer Metastasis Rev* 22: 259-269.

Chiang KP, Niessen S, Saghatelian A & Cravatt BF. (2006) An enzyme that regulates ether lipid signaling pathways in cancer annotated by multidimensional profiling. *Chem Biol* 13: 1041-1050.

Chong S, Lee KS, Chung MJ, Han J, Kwon OJ & Kim TS. (2006) Neuroendocrine tumors of the lung: clinical, pathologic, and imaging findings. *Radiographics* 26: 41-57; discussion 57-48.

Collaud S, Wiedl T, Cattaneo E, Soltermann A, Hillinger S, Weder W & Arni S. (2010) Laser-capture microdissection impairs activity-based protein profiles for serine hydrolase in human lung adenocarcinoma. *J Biomol Tech* 21: 25-28.

Collins GT, Brim RL, Narasimhan D, Ko MC, Sunahara RK, Zhan CG & Woods JH. (2009) Cocaine esterase prevents cocaine-induced toxicity and the ongoing intravenous self-administration of cocaine in rats. *J Pharmacol Exp Ther* 331: 445-455.

Corfield AP, Wagner SA, Clamp JR, Kriaris MS & Hoskins LC. (1992) Mucin degradation in the human colon: production of sialidase, sialate O-acetylesterase, N-acetylneuraminate lyase, arylesterase, and glycosulfatase activities by strains of fecal bacteria. *Infect Immun* 60: 3971-3978.

Cravatt BF, Wright AT & Kozarich JW. (2008) Activity-based protein profiling: from enzyme chemistry to proteomic chemistry. *Annu Rev Biochem* 77: 383-414.

Creighton TE. (1993) *Proteins: Structure and Molecular Properties,* New York: Freeman.

Cullen SP, Brunet M & Martin SJ. (2010) Granzymes in cancer and immunity. *Cell Death Differ* 17: 616-623.

Cunningham D, Swartzlander D, Liyanarachchi S, Davuluri RV & Herman GE. (2005) Changes in gene expression associated with loss of function of the NSDHL sterol dehydrogenase in mouse embryonic fibroblasts. *J Lipid Res* 46: 1150-1162.

Dammann R, Li C, Yoon JH, Chin PL, Bates S & Pfeifer GP. (2000) Epigenetic inactivation of a RAS association domain family protein from the lung tumour suppressor locus 3p21.3. *Nat Genet* 25: 315-319.

Dekker FJ, Rocks O, Vartak N, Menninger S, Hedberg C, Balamurugan R, Wetzel S, Renner S, Gerauer M, Scholermann B, Rusch M, Kramer JW, Rauh D, Coates GW, Brunsveld L, Bastiaens PI & Waldmann H. (2010) Small-molecule inhibition of APT1 affects Ras localization and signaling. *Nat Chem Biol* 6: 449-456.

Desiere F, Deutsch EW, Nesvizhskii AI, Mallick P, King NL, Eng JK, Aderem A, Boyle R, Brunner E, Donohoe S, Fausto N, Hafen E, Hood L, Katze MG, Kennedy KA, Kregenow F, Lee H, Lin B, Martin D, Ranish JA, Rawlings DJ, Samelson LE, Shiio Y, Watts JD, Wollscheid B, Wright ME, Yan W,

Yang L, Yi EC, Zhang H & Aebersold R. (2005) Integration with the human genome of peptide sequences obtained by high-throughput mass spectrometry. *Genome Biol* 6: R9.

Dijkstra HP, Sprong H, Aerts BN, Kruithof CA, Egmond MR & Klein Gebbink RJ. (2008) Selective and diagnostic labelling of serine hydrolases with reactive phosphonate inhibitors. *Org Biomol Chem* 6: 523-531.

Domon B & Aebersold R. (2010) Options and considerations when selecting a quantitative proteomics strategy. *Nat Biotechnol* 28: 710-721.

Duan L, Motchoulski N, Danzer B, Davidovich I, Shariat-Madar Z & Levenson VV. (2011) Prolylcarboxypeptidase regulates proliferation, autophagy, and resistance to 4-hydroxytamoxifen-induced cytotoxicity in estrogen receptor-positive breast cancer cells. *J Biol Chem* 286: 2864-2876.

Dusmet MaG, Peter. (2008) Staging, classification, and prognosis. In: Hanson H (ed) *Textbook of lung cancer.* second ed. London, UK: Informa Healthcare, 97-122.

Eiberg H & Mohr J. (1986) Identity of the polymorphisms for esterase D and S-formylglutathione hydrolase in red blood cells. *Hum Genet* 74: 174-175.

Eisen MB, Spellman PT, Brown PO & Botstein D. (1998) Cluster analysis and display of genome-wide expression patterns. *Proc Natl Acad Sci U S A* 95: 14863-14868.

Ekici OD, Paetzel M & Dalbey RE. (2008) Unconventional serine proteases: variations on the catalytic Ser/His/Asp triad configuration. *Protein Sci* 17: 2023-2037.

Forner F, Foster LJ, Campanaro S, Valle G & Mann M. (2006) Quantitative proteomic comparison of rat mitochondria from muscle, heart, and liver. *Mol Cell Proteomics* 5: 608-619.

Friend SH, Bernards R, Rogelj S, Weinberg RA, Rapaport JM, Albert DM & Dryja TP. (1986) A human DNA segment with properties of the gene that predisposes to retinoblastoma and osteosarcoma. *Nature* 323: 643-646.

Garcia M, Jemal, A, Ward, EM, Center, MM, Hao, Y, Siegel, RL, Thun, MJ. (2007) Global Cancer Facts & Figures 2007. Atlanta, GA: American Cancer Society.

Glynn P. (1999) Neuropathy target esterase. *Biochem J* 344 Pt 3: 625-631.

Gonzalez CF, Proudfoot M, Brown G, Korniyenko Y, Mori H, Savchenko AV & Yakunin AF. (2006) Molecular basis of formaldehyde detoxification. Characterization of two S-formylglutathione hydrolases from Escherichia coli, FrmB and YeiG. *J Biol Chem* 281: 14514-14522.

Grossmann J, Roschitzki B, Panse C, Fortes C, Barkow-Oesterreicher S, Rutishauser D & Schlapbach R. (2010) Implementation and evaluation of relative and absolute quantification in shotgun proteomics with label-free methods. *J Proteomics* 73: 1740-1746.

Gygi SP, Rist B, Gerber SA, Turecek F, Gelb MH & Aebersold R. (1999a) Quantitative analysis of complex protein mixtures using isotope-coded affinity tags. *Nat Biotechnol* 17: 994-999.

Gygi SP, Rochon Y, Franza BR & Aebersold R. (1999b) Correlation between protein and mRNA abundance in yeast. *Mol Cell Biol* 19: 1720-1730.

Hanahan D & Weinberg RA. (2000) The hallmarks of cancer. *Cell* 100: 57-70.

Haugen AaM, Steen. (2008) Etiology of lung cancer. In: Hanson H (ed) *Textbook of lung cancer.* second ed. London, UK: Informa Healthcare, 61-74.

Haws C, Finkbeiner WE, Widdicombe JH & Wine JJ. (1994) CFTR in Calu-3 human airway cells: channel properties and role in cAMP-activated Cl-conductance. *Am J Physiol* 266: L502-512.

Herbst RS, Heymach JV & Lippman SM. (2008) Lung cancer. *N Engl J Med* 359: 1367-1380.

Hernandez J & Thompson IM. (2004) Prostate-specific antigen: a review of the validation of the most commonly used cancer biomarker. *Cancer* 101: 894-904.

Higgins DG & Sharp PM. (1988) CLUSTAL: a package for performing multiple sequence alignment on a microcomputer. *Gene* 73: 237-244.

Higson AP, Ferguson MAJ & Nikolaev AV. (1999) Synthesis of 6-N-Biotinylaminohexyl Isopropyl Phosphorofluoridate: A Potent Tool for the Inhibition/Isolation of Serine Esterases and Proteases. *Synthesis*: 407-409.

Hiraiwa M. (1999) Cathepsin A/protective protein: an unusual lysosomal multifunctional protein. *Cell Mol Life Sci* 56: 894-907.

Hirano T, Kishi M, Sugimoto H, Taguchi R, Obinata H, Ohshima N, Tatei K & Izumi T. (2009) Thioesterase activity and subcellular localization of acylprotein thioesterase 1/lysophospholipase 1. *Biochim Biophys Acta* 1791: 797-805.

Hong WKaT, A.S. (2008) *Lung Carcinoma: Tumors of the Lungs.*

Hopkins AL & Groom CR. (2002) The druggable genome. *Nat Rev Drug Discov* 1: 727-730.

Hu L, Ye M, Jiang X, Feng S & Zou H. (2007) Advances in hyphenated analytical techniques for shotgun proteome and peptidome analysis--a review. *Anal Chim Acta* 598: 193-204.

Hu Q, Noll RJ, Li H, Makarov A, Hardman M & Graham Cooks R. (2005) The Orbitrap: a new mass spectrometer. *J Mass Spectrom* 40: 430-443.

Huang F, Reeves K, Han X, Fairchild C, Platero S, Wong TW, Lee F, Shaw P & Clark E. (2007) Identification of candidate molecular markers predicting sensitivity in solid tumors to dasatinib: rationale for patient selection. *Cancer Res* 67: 2226-2238.

Huang Y, Wang S & Kelly T. (2004) Seprase promotes rapid tumor growth and increased microvessel density in a mouse model of human breast cancer. *Cancer Res* 64: 2712-2716.

Hung KE & Yu KH. (2010) Proteomic approaches to cancer biomarkers. *Gastroenterology* 138: 46-51 e41.

Iggo R, Gatter K, Bartek J, Lane D & Harris AL. (1990) Increased expression of mutant forms of p53 oncogene in primary lung cancer. *Lancet* 335: 675-679.

Inge LJ, Coon KD, Smith MA & Bremner RM. (2009) Expression of LKB1 tumor suppressor in non-small cell lung cancer determines sensitivity to 2-deoxyglucose. *J Thorac Cardiovasc Surg* 137: 580-586.

Janssen-Heijnen ML & Coebergh JW. (2003) The changing epidemiology of lung cancer in Europe. *Lung Cancer* 41: 245-258.

Jessani N, Liu Y, Humphrey M & Cravatt BF. (2002) Enzyme activity profiles of the secreted and membrane proteome that depict cancer cell invasiveness. *Proc Natl Acad Sci U S A* 99: 10335-10340.

Jessani N, Niessen S, Wei BQ, Nicolau M, Humphrey M, Ji Y, Han W, Noh DY, Yates JR, 3rd, Jeffrey SS & Cravatt BF. (2005) A streamlined platform for high-content functional proteomics of primary human specimens. *Nat Methods* 2: 691-697.

Jewell C, Bennett P, Mutch E, Ackermann C & Williams FM. (2007) Inter-individual variability in esterases in human liver. *Biochem Pharmacol* 74: 932-939.

Ji H, Ramsey MR, Hayes DN, Fan C, McNamara K, Kozlowski P, Torrice C, Wu MC, Shimamura T, Perera SA, Liang MC, Cai D, Naumov GN, Bao L, Contreras CM, Li D, Chen L, Krishnamurthy J, Koivunen J, Chirieac LR, Padera RF, Bronson RT, Lindeman NI, Christiani DC, Lin X, Shapiro GI, Janne PA, Johnson BE, Meyerson M, Kwiatkowski DJ, Castrillon DH, Bardeesy N, Sharpless NE & Wong KK. (2007) LKB1 modulates lung cancer differentiation and metastasis. *Nature* 448: 807-810.

Kaira K, Oriuchi N, Imai H, Shimizu K, Yanagitani N, Sunaga N, Hisada T, Tanaka S, Ishizuka T, Kanai Y, Endou H, Nakajima T & Mori M. (2008) Prognostic significance of L-type amino acid transporter 1 expression in resectable stage I-III nonsmall cell lung cancer. *Br J Cancer* 98: 742-748.

Karuman P, Gozani O, Odze RD, Zhou XC, Zhu H, Shaw R, Brien TP, Bozzuto CD, Ooi D, Cantley LC & Yuan J. (2001) The Peutz-Jegher gene product LKB1 is a mediator of p53-dependent cell death. *Mol Cell* 7: 1307-1319.

Khazaie K, Blatner NR, Khan MW, Gounari F, Gounaris E, Dennis K, Bonertz A, Tsai FN, Strouch MJ, Cheon E, Phillips JD, Beckhove P & Bentrem DJ. (2011) The significant role of mast cells in cancer. *Cancer Metastasis Rev.*

Kilpinen S, Autio R, Ojala K, Iljin K, Bucher E, Sara H, Pisto T, Saarela M, Skotheim RI, Bjorkman M, Mpindi JP, Haapa-Paananen S, Vainio P, Edgren H, Wolf M, Astola J, Nees M, Hautaniemi S & Kallioniemi O. (2008) Systematic bioinformatic analysis of expression levels of 17,330 human genes across 9,783 samples from 175 types of healthy and pathological tissues. *Genome Biol* 9: R139.

Kok K, Osinga J, Carritt B, Davis MB, van der Hout AH, van der Veen AY, Landsvater RM, de Leij LF, Berendsen HH, Postmus PE & et al. (1987) Deletion of a DNA sequence at the chromosomal region 3p21 in all major types of lung cancer. *Nature* 330: 578-581.

Kollmann K, Damme M, Deuschl F, Kahle J, D'Hooge R, Lullmann-Rauch R & Lubke T. (2009) Molecular characterization and gene disruption of mouse lysosomal putative serine carboxypeptidase 1. *FEBS J* 276: 1356-1369.

Kollmann K, Mutenda KE, Balleininger M, Eckermann E, von Figura K, Schmidt B & Lubke T. (2005) Identification of novel lysosomal matrix proteins by proteome analysis. *Proteomics* 5: 3966-3978.

Korst R. (2008) Treatment of non-small cell lung cancer. In: Hanson H (ed) *Textbook of lung cancer.* second ed. London, UK: Informa Healthcare, 123-135.

Kuhajda FP. (2000) Fatty-acid synthase and human cancer: new perspectives on its role in tumor biology. *Nutrition* 16: 202-208.

Lange V, Picotti P, Domon B & Aebersold R. (2008) Selected reaction monitoring for quantitative proteomics: a tutorial. *Mol Syst Biol* 4: 222.

Larrinaga G, Perez I, Blanco L, Lopez JI, Andres L, Etxezarraga C, Santaolalla F, Zabala A, Varona A & Irazusta J. (2010) Increased prolyl endopeptidase activity in human neoplasia. *Regul Pept* 163: 102-106.

Leung D, Hardouin C, Boger DL & Cravatt BF. (2003) Discovering potent and selective reversible inhibitors of enzymes in complex proteomes. *Nat Biotechnol* 21: 687-691.

Li F, Fei X, Xu J & Ji C. (2009) An unannotated alpha/beta hydrolase superfamily member, ABHD6 differentially expressed among cancer cell lines. *Mol Biol Rep* 36: 691-696.

Li XJ, Pedrioli PG, Eng J, Martin D, Yi EC, Lee H & Aebersold R. (2004) A tool to visualize and evaluate data obtained by liquid chromatography-electrospray ionization-mass spectrometry. *Anal Chem* 76: 3856-3860.

Liu H, Sadygov RG & Yates JR, 3rd. (2004) A model for random sampling and estimation of relative protein abundance in shotgun proteomics. *Anal Chem* 76: 4193-4201.

Liu J, Hamza A & Zhan CG. (2009a) Fundamental reaction mechanism and free energy profile for (-)-cocaine hydrolysis catalyzed by cocaine esterase. *J Am Chem Soc* 131: 11964-11975.

Liu Y, Patricelli MP & Cravatt BF. (1999) Activity-based protein profiling: the serine hydrolases. *Proc Natl Acad Sci U S A* 96: 14694-14699.

Liu YF, Xiao ZQ, Li MX, Li MY, Zhang PF, Li C, Li F, Chen YH, Yi H, Yao HX & Chen ZC. (2009b) Quantitative proteome analysis reveals annexin A3 as a novel biomarker in lung adenocarcinoma. *J Pathol* 217: 54-64.

Loong TW. (2003) Understanding sensitivity and specificity with the right side of the brain. *BMJ* 327: 716-719.

Maes MB, Lambeir AM, Gilany K, Senten K, Van der Veken P, Leiting B, Augustyns K, Scharpe S & De Meester I. (2005) Kinetic investigation of human dipeptidyl peptidase II (DPPII)-mediated hydrolysis of dipeptide derivatives and its identification as quiescent cell proline dipeptidase (QPP)/dipeptidyl peptidase 7 (DPP7). *Biochem J* 386: 315-324.

Mahoney JA, Ntolosi B, DaSilva RP, Gordon S & McKnight AJ. (2001) Cloning and characterization of CPVL, a novel serine carboxypeptidase, from human macrophages. *Genomics* 72: 243-251.

Malmstrom J, Beck M, Schmidt A, Lange V, Deutsch EW & Aebersold R. (2009) Proteome-wide cellular protein concentrations of the human pathogen Leptospira interrogans. *Nature* 460: 762-765.

Mao L, Lee JS, Kurie JM, Fan YH, Lippman SM, Lee JJ, Ro JY, Broxson A, Yu R, Morice RC, Kemp BL, Khuri FR, Walsh GL, Hittelman WN & Hong WK. (1997) Clonal genetic alterations in the lungs of current and former smokers. *J Natl Cancer Inst* 89: 857-862.

Marchler-Bauer A, Anderson JB, Chitsaz F, Derbyshire MK, DeWeese-Scott C, Fong JH, Geer LY, Geer RC, Gonzales NR, Gwadz M, He S, Hurwitz DI, Jackson JD, Ke Z, Lanczycki CJ, Liebert CA, Liu C, Lu F, Lu S, Marchler GH, Mullokandov M, Song JS, Tasneem A, Thanki N, Yamashita RA, Zhang D, Zhang N & Bryant SH. (2009) CDD: specific functional annotation with the Conserved Domain Database. *Nucleic Acids Res* 37: D205-210.

Marchler-Bauer A, Anderson JB, Derbyshire MK, DeWeese-Scott C, Gonzales NR, Gwadz M, Hao L, He S, Hurwitz DI, Jackson JD, Ke Z, Krylov D, Lanczycki CJ, Liebert CA, Liu C, Lu F, Lu S, Marchler GH, Mullokandov

M, Song JS, Thanki N, Yamashita RA, Yin JJ, Zhang D & Bryant SH. (2007) CDD: a conserved domain database for interactive domain family analysis. *Nucleic Acids Res* 35: D237-240.

Marchler-Bauer A, Panchenko AR, Shoemaker BA, Thiessen PA, Geer LY & Bryant SH. (2002) CDD: a database of conserved domain alignments with links to domain three-dimensional structure. *Nucleic Acids Res* 30: 281-283.

Marian AJ. (2009) Nature's genetic gradients and the clinical phenotype. *Circ Cardiovasc Genet* 2: 537-539.

Marrs WR, Blankman JL, Horne EA, Thomazeau A, Lin YH, Coy J, Bodor AL, Muccioli GG, Hu SS, Woodruff G, Fung S, Lafourcade M, Alexander JP, Long JZ, Li W, Xu C, Moller T, Mackie K, Manzoni OJ, Cravatt BF & Stella N. (2010) The serine hydrolase ABHD6 controls the accumulation and efficacy of 2-AG at cannabinoid receptors. *Nat Neurosci* 13: 951-957.

Mele DA, Bista P, Baez DV & Huber BT. (2009) Dipeptidyl peptidase 2 is an essential survival factor in the regulation of cell quiescence. *Cell Cycle* 8: 2425-2434.

Merla G, Ucla C, Guipponi M & Reymond A. (2002) Identification of additional transcripts in the Williams-Beuren syndrome critical region. *Hum Genet* 110: 429-438.

Merlo A, Gabrielson E, Askin F & Sidransky D. (1994) Frequent loss of chromosome 9 in human primary non-small cell lung cancer. *Cancer Res* 54: 640-642.

Meyer AM, Dwyer-Nield LD, Hurteau GJ, Keith RL, O'Leary E, You M, Bonventre JV, Nemenoff RA & Malkinson AM. (2004) Decreased lung tumorigenesis in mice genetically deficient in cytosolic phospholipase A2. *Carcinogenesis* 25: 1517-1524.

Mills GB & Moolenaar WH. (2003) The emerging role of lysophosphatidic acid in cancer. *Nat Rev Cancer* 3: 582-591.

Minna J. (2005) Neoplasms of the Lung. In: Kasper DL (ed) *Harrison`s Principles of Internal Medicine.* 16 ed., 506-515.

Mitta M, Ohnogi H, Mizutani S, Sakiyama F, Kato I & Tsunasawa S. (1996) The nucleotide sequence of human acylamino acid-releasing enzyme. *DNA Res* 3: 31-35.

Mountain CF & Dresler CM. (1997) Regional lymph node classification for lung cancer staging. *Chest* 111: 1718-1723.

Mu D, Hsu DS & Sancar A. (1996) Reaction mechanism of human DNA repair excision nuclease. *J Biol Chem* 271: 8285-8294.

Mullon JaO, Eric. (2008) Clinical diagnosis and basic evaluation. In: Hanson H (ed) *Textbook of lung cancer.* London, UK: Informa Healthcare, 75-96.

Nomura DK, Dix MM & Cravatt BF. (2010a) Activity-based protein profiling for biochemical pathway discovery in cancer. *Nat Rev Cancer* 10: 630-638.

Nomura DK, Leung D, Chiang KP, Quistad GB, Cravatt BF & Casida JE. (2005) A brain detoxifying enzyme for organophosphorus nerve poisons. *Proc Natl Acad Sci U S A* 102: 6195-6200.

Nomura DK, Long JZ, Niessen S, Hoover HS, Ng SW & Cravatt BF. (2010b) Monoacylglycerol lipase regulates a fatty acid network that promotes cancer pathogenesis. *Cell* 140: 49-61.

Ohtsuka K, Inoue S, Kameyama M, Kanetoshi A, Fujimoto T, Takaoka K, Araya Y & Shida A. (2003) Intracellular conversion of irinotecan to its active form, SN-38, by native carboxylesterase in human non-small cell lung cancer. *Lung Cancer* 41: 187-198.

Olaussen KA, Dunant A, Fouret P, Brambilla E, Andre F, Haddad V, Taranchon E, Filipits M, Pirker R, Popper HH, Stahel R, Sabatier L, Pignon JP, Tursz T, Le Chevalier T & Soria JC. (2006) DNA repair by ERCC1 in non-small-cell lung cancer and cisplatin-based adjuvant chemotherapy. *N Engl J Med* 355: 983-991.

Ollis DL, Cheah E, Cygler M, Dijkstra B, Frolow F, Franken SM, Harel M, Remington SJ, Silman I, Schrag J & et al. (1992) The alpha/beta hydrolase fold. *Protein Eng* 5: 197-211.

Olsen C & Wagtmann N. (2002) Identification and characterization of human DPP9, a novel homologue of dipeptidyl peptidase IV. *Gene* 299: 185-193.

Orita H, Coulter J, Lemmon C, Tully E, Vadlamudi A, Medghalchi SM, Kuhajda FP & Gabrielson E. (2007) Selective inhibition of fatty acid synthase for lung cancer treatment. *Clin Cancer Res* 13: 7139-7145.

Otterson GA, Kratzke RA, Coxon A, Kim YW & Kaye FJ. (1994) Absence of p16INK4 protein is restricted to the subset of lung cancer lines that retains wildtype RB. *Oncogene* 9: 3375-3378.

Pabarcus MK & Casida JE. (2005) Cloning, expression, and catalytic triad of recombinant arylformamidase. *Protein Expr Purif* 44: 39-44.

Paez JG, Janne PA, Lee JC, Tracy S, Greulich H, Gabriel S, Herman P, Kaye FJ, Lindeman N, Boggon TJ, Naoki K, Sasaki H, Fujii Y, Eck MJ, Sellers WR, Johnson BE & Meyerson M. (2004) EGFR mutations in lung cancer: correlation with clinical response to gefitinib therapy. *Science* 304: 1497-1500.

Pao W, Miller V, Zakowski M, Doherty J, Politi K, Sarkaria I, Singh B, Heelan R, Rusch V, Fulton L, Mardis E, Kupfer D, Wilson R, Kris M & Varmus H. (2004) EGF receptor gene mutations are common in lung cancers from "never smokers" and are associated with sensitivity of tumors to gefitinib and erlotinib. *Proc Natl Acad Sci U S A* 101: 13306-13311.

Pao W, Miller VA, Politi KA, Riely GJ, Somwar R, Zakowski MF, Kris MG & Varmus H. (2005) Acquired resistance of lung adenocarcinomas to gefitinib or erlotinib is associated with a second mutation in the EGFR kinase domain. *PLoS Med* 2: e73.

Parkin DM, Pisani P & Ferlay J. (1999) Global cancer statistics. *CA Cancer J Clin* 49: 33-64, 31.

Parsam VL, Kannabiran C, Honavar S, Vemuganti GK & Ali MJ. (2009) A comprehensive, sensitive and economical approach for the detection of mutations in the RB1 gene in retinoblastoma. *J Genet* 88: 517-527.

Patricelli MP & Cravatt BF. (1999) Fatty acid amide hydrolase competitively degrades bioactive amides and esters through a nonconventional catalytic mechanism. *Biochemistry* 38: 14125-14130.

Payne V & Kam PC. (2004) Mast cell tryptase: a review of its physiology and clinical significance. *Anaesthesia* 59: 695-703.

Peng CYJ, Lee KL & Ingersoll GM. (2002) An Introduction to Logistic Regression Analysis and Reporting. *The Journal of Educational Research* 96.

Picotti P, Lam H, Campbell D, Deutsch EW, Mirzaei H, Ranish J, Domon B & Aebersold R. (2008) A database of mass spectrometric assays for the yeast proteome. *Nat Methods* 5: 913-914.

Poulsen T, Poulsen, Hans and Pappot, Helle. (2008) Molecular biology of lung cancer. In: Hanson H (ed) *Textbook of lung cancer.* second ed. London, UK: Informa Healthcare, 20-34.

Proctor RN. (2001) Tobacco and the global lung cancer epidemic. *Nat Rev Cancer* 1: 82-86.

Quistad GB & Casida JE. (2004) Lysophospholipase inhibition by organophosphorus toxicants. *Toxicol Appl Pharmacol* 196: 319-326.

Raponi M, Zhang Y, Yu J, Chen G, Lee G, Taylor JM, Macdonald J, Thomas D, Moskaluk C, Wang Y & Beer DG. (2006) Gene expression signatures for predicting prognosis of squamous cell and adenocarcinomas of the lung. *Cancer Res* 66: 7466-7472.

Ren S, Sakai K & Schwartz LB. (1998) Regulation of human mast cell beta-tryptase: conversion of inactive monomer to active tetramer at acid pH. *J Immunol* 160: 4561-4569.

Rodriguez-Pineiro AM, Blanco-Prieto S, Sanchez-Otero N, Rodriguez-Berrocal FJ & de la Cadena MP. (2010) On the identification of biomarkers for non-small cell lung cancer in serum and pleural effusion. *J Proteomics* 73: 1511-1522.

Russ AP & Lampel S. (2005) The druggable genome: an update. *Drug Discov Today* 10: 1607-1610.

Saviozzi S, Cordero F, Lo Iacono M, Novello S, Scagliotti GV & Calogero RA. (2006) Selection of suitable reference genes for accurate normalization of gene expression profile studies in non-small cell lung cancer. *BMC Cancer* 6: 200.

Sawyers CL. (2008) The cancer biomarker problem. *Nature* 452: 548-552.

Schopfer LM, Voelker T, Bartels CF, Thompson CM & Lockridge O. (2005) Reaction kinetics of biotinylated organophosphorus toxicant, FP-biotin, with human acetylcholinesterase and human butyrylcholinesterase. *Chem Res Toxicol* 18: 747-754.

Schubert C. (2009) The genomic basis of the Williams-Beuren syndrome. *Cell Mol Life Sci* 66: 1178-1197.

Scott WJ, Howington J, Feigenberg S, Movsas B & Pisters K. (2007) Treatment of non-small cell lung cancer stage I and stage II: ACCP evidence-based clinical practice guidelines (2nd edition). *Chest* 132: 234S-242S.

Sculier J-P, Meert, Anne-Pascal, Paesmans, Marianne and Berghmans, Thierry. (2008) Textbook of lung cancer. In: Hanson H (ed) *Textbook of lung cancer.* second ed. London, UK: Informa Healthcare, 236-246.

Selvaggi G, Novello S, Torri V, Leonardo E, De Giuli P, Borasio P, Mossetti C, Ardissone F, Lausi P & Scagliotti GV. (2004) Epidermal growth factor receptor overexpression correlates with a poor prognosis in completely resected non-small-cell lung cancer. *Ann Oncol* 15: 28-32.

Selvaggi GaS, Giorgio Vittorio. (2008) The future. In: Hanson H (ed) *Textbook of lung cancer.* second ed. London, UK: Informa Healthcare, 264-274.

Sequist LV, Bell DW, Lynch TJ & Haber DA. (2007) Molecular predictors of response to epidermal growth factor receptor antagonists in non-small-cell lung cancer. *J Clin Oncol* 25: 587-595.

Sharma SV, Bell DW, Settleman J & Haber DA. (2007) Epidermal growth factor receptor mutations in lung cancer. *Nat Rev Cancer* 7: 169-181.

Shayman JA, Kelly R, Kollmeyer J, He Y & Abe A. (2011) Group XV phospholipase A, a lysosomal phospholipase A. *Prog Lipid Res* 50: 1-13.

Shields DJ, Niessen S, Murphy EA, Mielgo A, Desgrosellier JS, Lau SK, Barnes LA, Lesperance J, Bouvet M, Tarin D, Cravatt BF & Cheresh DA. (2010) RBBP9: a tumor-associated serine hydrolase activity required for pancreatic neoplasia. *Proc Natl Acad Sci U S A* 107: 2189-2194.

Sieber SA & Cravatt BF. (2006) Analytical platforms for activity-based protein profiling--exploiting the versatility of chemistry for functional proteomics. *Chem Commun (Camb)*: 2311-2319.

Sieber SA, Mondala TS, Head SR & Cravatt BF. (2004) Microarray platform for profiling enzyme activities in complex proteomes. *J Am Chem Soc* 126: 15640-15641.

Sieber SA, Niessen S, Hoover HS & Cravatt BF. (2006) Proteomic profiling of metalloprotease activities with cocktails of active-site probes. *Nat Chem Biol* 2: 274-281.

Silva JC, Gorenstein MV, Li GZ, Vissers JP & Geromanos SJ. (2006) Absolute quantification of proteins by LCMSE: a virtue of parallel MS acquisition. *Mol Cell Proteomics* 5: 144-156.

Simon GM & Cravatt BF. (2010) Activity-based proteomics of enzyme superfamilies: serine hydrolases as a case study. *J Biol Chem* 285: 11051-11055.

Sparkes RS, Murphree AL, Lingua RW, Sparkes MC, Field LL, Funderburk SJ & Benedict WF. (1983) Gene for hereditary retinoblastoma assigned to human chromosome 13 by linkage to esterase D. *Science* 219: 971-973.

Steu S, Baucamp M, von Dach G, Bawohl M, Dettwiler S, Storz M, Moch H & Schraml P. (2008) A procedure for tissue freezing and processing applicable to both intra-operative frozen section diagnosis and tissue banking in surgical pathology. *Virchows Arch* 452: 305-312.

Sun S, Schiller JH & Gazdar AF. (2007) Lung cancer in never smokers--a different disease. *Nat Rev Cancer* 7: 778-790.

Suster SaM, C. (2007) Tumors of the Lung and Pleura. In: Fan IDaF (ed) *Cancer Grading Manual*. New York: Springer, 23-31.

Tang X, Shigematsu H, Bekele BN, Roth JA, Minna JD, Hong WK, Gazdar AF & Wistuba, II. (2005) EGFR tyrosine kinase domain mutations are detected in histologically normal respiratory epithelium in lung cancer patients. *Cancer Res* 65: 7568-7572.

Tennis M, Krishnan S, Bonner M, Ambrosone CB, Vena JE, Moysich K, Swede H, McCann S, Hall P, Shields PG & Freudenheim JL. (2006) p53 Mutation analysis in breast tumors by a DNA microarray method. *Cancer Epidemiol Biomarkers Prev* 15: 80-85.

Thompson JD, Higgins DG & Gibson TJ. (1994) CLUSTAL W: improving the sensitivity of progressive multiple sequence alignment through sequence weighting, position-specific gap penalties and weight matrix choice. *Nucleic Acids Res* 22: 4673-4680.

Thors L, Bergh A, Persson E, Hammarsten P, Stattin P, Egevad L, Granfors T & Fowler CJ. (2010) Fatty acid amide hydrolase in prostate cancer: association with disease severity and outcome, CB1 receptor expression and regulation by IL-4. *PLoS One* 5: e12275.

Uhlen M, Bjorling E, Agaton C, Szigyarto CA, Amini B, Andersen E, Andersson AC, Angelidou P, Asplund A, Asplund C, Berglund L, Bergstrom K, Brumer H, Cerjan D, Ekstrom M, Elobeid A, Eriksson C, Fagerberg L, Falk R, Fall J, Forsberg M, Bjorklund MG, Gumbel K, Halimi A, Hallin I, Hamsten C, Hansson M, Hedhammar M, Hercules G, Kampf C, Larsson K, Lindskog M, Lodewyckx W, Lund J, Lundeberg J, Magnusson K, Malm E, Nilsson P, Odling J, Oksvold P, Olsson I, Oster E, Ottosson J, Paavilainen L, Persson A, Rimini R, Rockberg J, Runeson M, Sivertsson A, Skollermo A, Steen J, Stenvall M, Sterky F, Stromberg S, Sundberg M, Tegel H, Tourle S, Wahlund E, Walden A, Wan J, Wernerus H, Westberg J, Wester K, Wrethagen U, Xu LL, Hober S & Ponten F. (2005) A human protein atlas for normal and cancer tissues based on antibody proteomics. *Mol Cell Proteomics* 4: 1920-1932.

van Straaten KE, Gonzalez CF, Valladares RB, Xu X, Savchenko AV & Sanders DA. (2009) The structure of a putative S-formylglutathione hydrolase from Agrobacterium tumefaciens. *Protein Sci* 18: 2196-2202.

Walsh CT. (1979) *Enzymatic Reaction Mechanisms,* New York: Freeman.

Washburn MP, Wolters D & Yates JR, 3rd. (2001) Large-scale analysis of the yeast proteome by multidimensional protein identification technology. *Nat Biotechnol* 19: 242-247.

Weerapana E, Speers AE & Cravatt BF. (2007) Tandem orthogonal proteolysis-activity-based protein profiling (TOP-ABPP)--a general method for mapping sites of probe modification in proteomes. *Nat Protoc* 2: 1414-1425.

Wesley UV, Tiwari S & Houghton AN. (2004) Role for dipeptidyl peptidase IV in tumor suppression of human non small cell lung carcinoma cells. *Int J Cancer* 109: 855-866.

Westra WH. (2000) Early glandular neoplasia of the lung. *Respir Res* 1: 163-169.

WHO. (1999) Tobacco or Health: A Global Status Report. Geneva: World Health Organization.

Wistuba, II, Behrens C, Virmani AK, Mele G, Milchgrub S, Girard L, Fondon JW, 3rd, Garner HR, McKay B, Latif F, Lerman MI, Lam S, Gazdar AF & Minna JD. (2000) High resolution chromosome 3p allelotyping of human lung cancer and preneoplastic/preinvasive bronchial epithelium reveals multiple, discontinuous sites of 3p allele loss and three regions of frequent breakpoints. *Cancer Res* 60: 1949-1960.

Wistuba, II, Mao L & Gazdar AF. (2002) Smoking molecular damage in bronchial epithelium. *Oncogene* 21: 7298-7306.

Wolters DA, Washburn MP & Yates JR, 3rd. (2001) An automated multidimensional protein identification technology for shotgun proteomics. *Anal Chem* 73: 5683-5690.

Wu D, Li Y, Song G, Zhang D, Shaw N & Liu ZJ. (2009) Crystal structure of human esterase D: a potential genetic marker of retinoblastoma. *FASEB J* 23: 1441-1446.

Yano S, Matsumori Y, Ikuta K, Ogino H, Doljinsuren T & Sone S. (2006) Current status and perspective of angiogenesis and antivascular therapeutic strategy: non-small cell lung cancer. *Int J Clin Oncol* 11: 73-81.

Yu DM, Wang XM, McCaughan GW & Gorrell MD. (2006) Extraenzymatic functions of the dipeptidyl peptidase IV-related proteins DP8 and DP9 in cell adhesion, migration and apoptosis. *FEBS J* 273: 2447-2460.

Zamble DB, Mu D, Reardon JT, Sancar A & Lippard SJ. (1996) Repair of cisplatin--DNA adducts by the mammalian excision nuclease. *Biochemistry* 35: 10004-10013.

Zeidman R, Jackson CS & Magee AI. (2009) Protein acyl thioesterases (Review). *Mol Membr Biol* 26: 32-41.

Zelvyte I, Stevens T, Westin U & Janciauskiene S. (2004) alpha1-antitrypsin and its C-terminal fragment attenuate effects of degranulated neutrophil-conditioned medium on lung cancer HCC cells, in vitro. *Cancer Cell Int* 4: 7.

Zhao T, Zhang H, Guo Y, Zhang Q, Hua G, Lu H, Hou Q, Liu H & Fan Z. (2007) Granzyme K cleaves the nucleosome assembly protein SET to induce single-stranded DNA nicks of target cells. *Cell Death Differ* 14: 489-499.

i want morebooks!

Buy your books fast and straightforward online - at one of world's fastest growing online book stores! Environmentally sound due to Print-on-Demand technologies.

Buy your books online at
www.get-morebooks.com

Kaufen Sie Ihre Bücher schnell und unkompliziert online – auf einer der am schnellsten wachsenden Buchhandelsplattformen weltweit! Dank Print-On-Demand umwelt- und ressourcenschonend produziert.

Bücher schneller online kaufen
www.morebooks.de

 VDM Verlagsservicegesellschaft mbH
Heinrich-Böcking-Str. 6-8   Telefon: +49 681 3720 174   info@vdm-vsg.de
D - 66121 Saarbrücken   Telefax: +49 681 3720 1749   www.vdm-vsg.de

Printed by Books on Demand GmbH, Norderstedt / Germany